Google Classroom 2020 and Beyond

A Beginner to Expert User Guide for Teachers and Students to Master the Use of Google Classroom for an Engaging, Virtual Distance Learning…With Graphical Illustrations

By

Nicholas Scott

Disclaimer

This publication is designed to provide competent and reliable information regarding the subject matter covered. However, the views expressed in this publication are those of the author, and should not be taken as expert instruction or official advice from Google. The reader is responsible for his or her own actions.

The author hereby disclaims any responsibility or liability whatsoever that is incurred from the use or application of the contents of this publication by the purchaser or reader.

Books By The Same Author

Apple Watch 5 Manual (2020 Edition)

How To Cancel Kindle Unlimited

How to Delete Books From Your Kindle Library

About The Author

Nicholas Scott is a tech geek fanatic, with over 15 years of experience working in the tech space, right in the heart of Silicon Valley. Nick, as friends fondly call him, holds a Bachelor's degree in Software Engineering from MIT and a Master's degree in Information Technology from Stanford.

His hobby and passion revolves around geeking about technologies, and writing tech and computer guides for just about any trends in information technology.

He is married to a lovely wife and has two kids.

Table of Contents

Introduction

In recent times, teachers and students have been moved to take learning beyond the four walls of the classroom by stepping into the digital world, a world that takes pride in itself over an unlimited reach of people, ideas, and skills.

Before now, only a few educators made use of Google Classroom, but today, it is being embraced with open arms by all stakeholders, showing them what they have been missing — a web-based platform with the capacity to bring down the invisible barriers of a classroom, bringing learners much closer to each other and their end goal.

Google Classroom offers a streamlined and easy way for teachers to stay organized in Google Drive, create, access, and share digital assignments and provide students with more efficient and effective feedback and communication system. If you are new to Google Classroom, think of this tool as a must-have communication hub and an assignment manager for you and your students.

At the end of this book, you will:

- Have a 360-degree understanding of what Google Classroom is all about and why this tool is a must-have for teachers, students, and school administrators.

- Know how to get started with accessing Google Classroom irrespective of your location.

- Be equipped with the practical know-how to create a class in Google Classroom as well as how you can add your students to your classroom without any hassle.

- Be able to create assignments for your students and how you can attach documents to your assignments using Google Docs.

- Be able to create timed quizzes for your students, add the quizzes to your classes as well as view and grade assignments submitted by your students.

- Know how to effectively engage and communicate with your students and keep them abreast of any activity or class updates.

- Know how students can join a teacher's class, locate and submit teacher's assignments, and engage with teachers and other students in the class.

- Be able to set up live video classes for your students using Google Meet as well as navigate your way around the usage of this tool.

- Know how to address the common Google Classroom questions, and problems asked and encountered by teachers.

- Gain insider knowledge on the Google Classroom tips and tricks that most teachers don't know, as well as some of the best apps and extensions that integrate with Google Classroom to enhance the teacher-student engagement experience.

And much more!

Are you ready? Let's begin!

Chapter 1

A-Z of Google Classroom

What is Google Classroom

Google Classroom, developed by Google, is a free web-based learning platform that allows teachers to create curriculums and conduct remote-online classes. Teachers can also share assignments seamlessly with students to complete, after which they are graded, without a need to have anything printed. It also serves as a means of information communication to students. For example, teachers can post upcoming assignments, post announcements, and even email both the students and their parents.

Google Classroom simplifies the collaboration between teacher-student by leveraging a variety of G Suite (Google Suite) services such as Google Drive, Google Docs, Google Forms, Google Sheets, Google Slides, and many other Google services to create and store assignments among others.

Who Can Use Google Classroom?

Anyone who has a Google account is eligible to make use of Google Classroom, especially organizations that make use of the G Suite services for educational purposes such as non-profit organizations, independent educators, schools, administrators, homeschoolers, and families. That being said, the primary users of Google Classroom, however, are teachers and students, which is going to be our core focus for the remaining pages of this book.

Main Features of Google Classroom

Assignments: Assignments can be created by teachers through the use of learning materials such as Google Form surveys, Youtube videos, or PDFs from Google Drive. These assignments can then be assigned to individual students or to all students. Also, teachers can immediately send out the assignment or schedule it for a particular day.

Customizable Grading System: A grading system can be chosen by teachers, as well as creating grade categories. For example, teachers can choose the following grading systems:

- **Total Points Grading**: Here, the total points earned by students are divided by the maximum points allowable.

- **Weighted by Category Grading**: Here, a weight is assigned to the grade categories, then, the average scores of the individual grade category are calculated and multiplied with the weight of the grade, giving you the total grade out of 100%.

- **No Overall Grade:** Here, the teacher may choose not to grade the students.

Virtual Discussions: Students can be invited by teachers to answer question-driven discussions and to respond to classmates. The "comments" section on google docs facilitates this two-way discussion by enabling teachers to provide feedback to students, which is a great way for students to be kept engaged, especially when they can't be seen. Some individual students can likewise be muted by teachers to prevent them from posting or commenting.

Announcements: Updates through announcements can be given to students by teachers. Announcements are posts without assignments. They are just notices for students, such as upcoming tests, deadlines, or any class-related work or activity. Announcements can be scheduled by teachers, and can likewise manage comments and replies made on the announcement post.

Live Classes: This is one of the new features enabled for Google Classroom, where real-time, virtual classes can be taken by teachers using Google Meet. Up to 250 people can be added to a Hangout call, and about 100,000 viewers can live-stream. These meetings can also be recorded such that students who missed the live session can later watch them.

Devices Supported By Google Classroom

Google Classroom supports browsers such as Google Chrome, Mozilla Firefox, Internet Explorer, and Safari. Also, the Google Classroom app is available for download on Android and Apple iOS mobile devices.

Benefits of Google Classroom

Google Classroom offers many benefits for teachers and students as a free online learning platform. Below are 10 reasons teachers should give Google Classroom a try.

Accessibility

Google Classroom is accessible from any computer or any mobile device that has an internet connection. Files uploaded by teachers and students are stored on Google Drive in a Classroom folder. Users can then access this folder anytime and anywhere without having to worry about their computers crashing and losing their files.

Exposure

Through Google Classroom, students are now exposed to an online learning system. Several college and university programs require that students enroll in one online class at the very least. Exposure to Google Classroom may help students in transitioning to other online learning management systems employed in higher education.

Paperless

With Google Classroom, teachers and students will no longer have a need to shuffle excessive amounts of paper since Classroom is completely paperless. When assignments and assessments are uploaded to Google Classroom, they are saved simultaneously to Google Drive. Assignments and assessments can then be completed by the students directly through Classroom, with their work saved to Drive. Missed work can be accessed by students due to absences as well as locate other resources they may require.

Q Search Drive

My Drive ▾

Name ↓   Owner

Classroom   me

Time Saver

Google Classroom is a big time saver. Since all the resources are saved in a single place, and the ability to access Classroom can be from anywhere, and from any mobile device or platform, teachers will have extra time in completing other tasks.

Communication

Built-in tools allow for communication with students and parents to happen in a breeze. With the in-built communication tool, and with teachers having full control over student comments and posts, both teachers and students can send emails, send private comments on assignments, post to the Stream, and can provide feedback on their work. Teachers can also communicate with parents via individual emails or via Classroom email summaries, including class announcements and due dates.

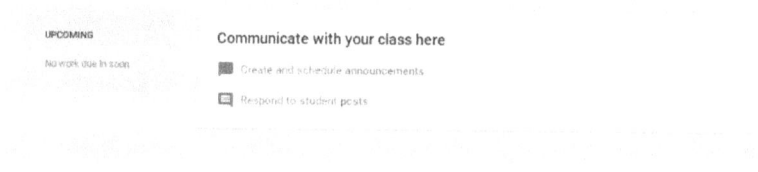

UPCOMING

No work due in soon

Communicate with your class here

Create and schedule announcements

Respond to student posts

Collaborate

Google Classroom provides several ways in which students can collaborate. For example, teachers can create group projects within Google Classroom and can champion online discussions between students. Students likewise can collaborate on Google Docs as shared by the teacher.

Engagement

Google Classroom provides several ways in making learning an interactive and collaborative activity. For instance, it allows teachers to differentiate assignments, create collaborative group assignments, and include web pages and videos into lessons.

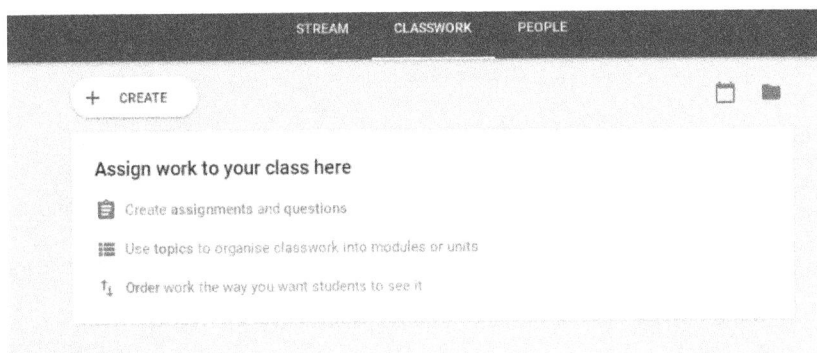

Differentiation

Through Google Classroom, teachers can easily differentiate instructions for learners by assigning lessons to the entire class, to individual students, or to groups of students, which usually takes a few simple steps when creating an assignment from the Classwork page.

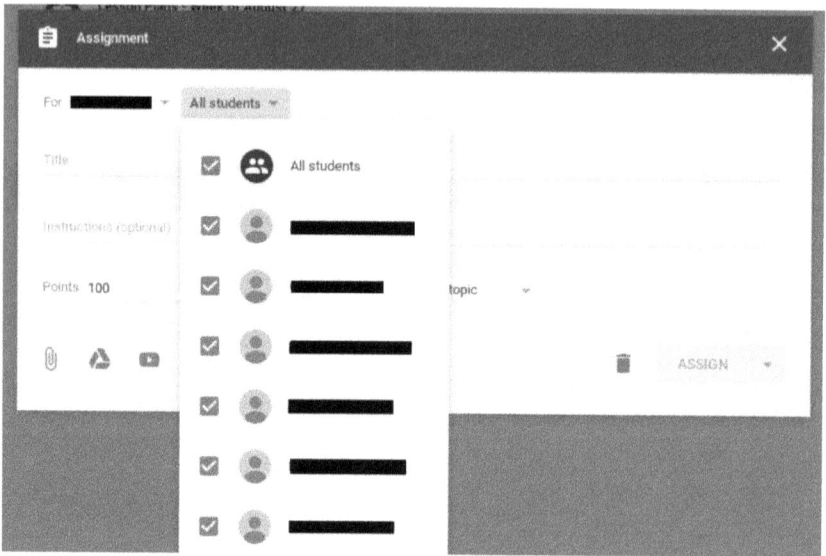

Feedback

Within the grading tool of Google Classroom, teachers can provide individual feedback on assignments to students. A comment bank can also be created within the grading tool for future use. Also, Google Classroom mobile app allows users to annotate work.

Data Analysis

To make learning meaningful, there is a need for teachers to analyze the data obtained from assessments to ensure that students understand the learning objectives. Data from assessments can be exported seamlessly into Google Sheets for sorting and analysis.

Top 10 Reasons Why Google Classroom is a Must-Have for Teachers

If you have been making use of Google Classroom for the last few years, you are most likely "sold" on why it is a good thing to have and use for online distance learning. However, if it is new to you, you would probably want to know why this app is useful both for teachers and students. Aside from the obvious benefits that Google Classroom provides, which points to why using this app is great, below are some other noteworthy things to consider:

- Google Classroom is user friendly. Compared to other Learning Management Systems that have over the past decade been popular, Google Classroom is user friendly, easy to use, and amazingly simple. Setting up a new classroom, for instance, takes less time and requires no expertise of some sort.

- Google Classroom allows you to communicate more efficiently. You input your student's email addresses just once, and the Classroom communication is completed. By entering students in the Classroom via email, you have an email group, a Google Calendar automatically created, and a discussion group. It is also easy to add and remove students from the class as the need arises.

- Google Classroom allows you to communicate more effectively. Perhaps more important than being easy to use and efficient, Google Classroom's communication tools are also very effective. Because Google Classroom is Cloud-based, losing of assignments by students can no longer occur, such that if a student is absent, communication remains seamless. There is also a parent notification feature that keeps parents notified and in the loop of what is going on in the Classroom.

- Google Classroom allows schools to be cost-effective and environmentally responsible. Personally, I am not totally sold on paperless learning, most especially for younger students. That being said, there is a real advantage for schools to become more cost-effective when it comes to printing. If each student has a device that is connected to the internet, every sheet of paper saved simply makes the school more efficient and responsible environmentally.

- Online learning is the new trendsetter for students and will continue to remain so in the future. For example, printing out the five-page essays by undergrads is no longer required by college campuses.

- With Google Classroom, teacher planning is seamless, easy, and assignments can be scheduled to take place in the future. For example, designated assignments can be scheduled to take place in July, on a Monday, and then close that Friday.

- With Google Classroom, classrooms can be used from semester to semester and from year to year. Just copying and pasting a classroom for the next group of students without making changes is terrible, but doing so does save some time in ensuring some things are in place, such as class syllabus, grading expectations, etc.

- Google Classroom constantly undergoes upgrades and improvements, which is by far one of its best selling points, well in my opinion. For example, if there is something that needs to be fixed or added, Google would actually listen and respond accordingly. The implication of this is that teachers will also have to continually learn how to use any new updates rolled out for Google Classroom, which isn't a bad thing either!

- With Google Classroom, the tendency of cheating among students is reduced since access to documents can be restricted to the teacher and the individual student, thereby preventing the temptation to copy each other.

- Google Classroom provides you with an easy view of student's submissions. It does this by displaying the number of students that have and have not submitted an assignment.

There are more reasons why Google Classroom is a must-have and a must-use for teachers and students alike. However, these ten excellent reasons in tandem with the benefits earlier discussed are why Google Classroom is worth giving a much closer look at.

Chapter 2

Getting Started with Google Classroom

Using Google Classroom – The Practical Approach

Teachers and students don't always have the same screen, do the same thing, or see the same thing in Google Classroom. There are slight differences in what students can see and do compared to teachers. Hence, getting a workflow down is important so that teachers and students know what to do. The procedures on how both teachers and students can navigate their way around Google Classroom is detailed below.

Teacher's Guide

Accessing Google Classroom

To start using Google Classroom, first, you need to ensure you have signed into your personal Google account – this applies if you want to use Google Classroom for personal use outside the school setting such as homeschooling. If you do not have one, kindly click this link creating a Google account or use this web address https://bit.ly/32L3RJv for a review lesson on how to open a google account. However, if Google Classroom is being used within a school setting, then you would have to sign in to a Google Classroom with a school account – this is also referred to as G Suite for Education account since this is the recommended account for a school setting. A school account is set up by the IT Admin department of accredited schools. Your school account could look like **you@yourschool.edu**.

After opening a Google account or after receiving your school account, Google Classroom can then be accessed by visiting classroom.google.com. Google Classroom also comes in a mobile app that you can download either on iPhone or on Android devices.

The mobile app is great for on-the-go access to your Classroom, but you will most likely have it easier to perform tasks such as grading and creating assignments using a computer.

Creating a Class

Upon opening Google Classroom, the first thing you need to do is to create a class. In the top-right corner, click the **Plus** icon then select **Create class**.

Create or join your first class!

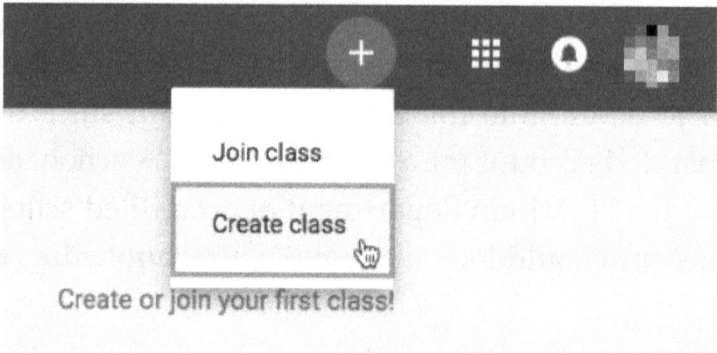

A dialog box would be displayed asking if you want to use Google Classroom at a school with students. If Google Classroom is to be used in actual classrooms, Google Classroom requires that schools use the **G Suite for Education** service for more privacy and security measures for teachers and students. However, if you are using Google Classroom for personal use, then there is no need to worry about this.

Using Classroom at a school with students?

If so, your school must sign up for a free G Suite for Education account before you can use Classroom. Learn More

G Suite for Education lets schools decide which Google services their students can use, and provides additional privacy and security protections that are important in a school setting. Students cannot use Google Classroom at a school with personal accounts.

☑ I've read and understand the above notice, and I'm not using Classroom at a school with students

GO BACK   CONTINUE

29

After accepting the terms and clicking continue, as shown above, you will then have to enter the **name** of your class. You can also enter a **Section, Subject,** and **Room** if you are using Google Classroom in a school setting, and would love to add this information. When you are done, click **Create**.

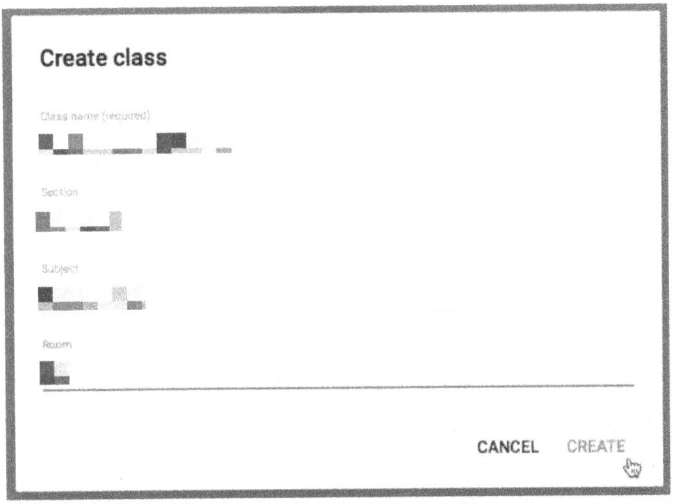

Personalizing a Class

Now, you should be in your Google Classroom that is similar to the image below. There is a large banner image at the top, which can be modified to personalize the class. This modification is also applied to the thumbnail class image for your students to help them distinguish your class from others, and also applies to you if you have multiple classes created.

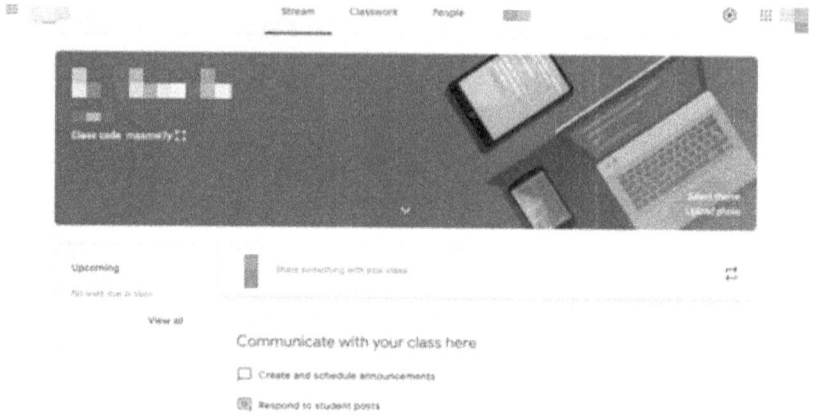

Within the top banner image are options located at the bottom right, which are "Select theme" or "Upload photo." The first option is the easiest because it offers several themes grouped by class type, including "General," "Math & Science," "English & History," "Arts," "Sports," and "Other." Select your choice and select "class theme" located at the bottom, and automatically, it will appear on your page. On the other hand, you can choose to upload your own photo. Just click the "upload photo" at the bottom right of the banner as shown above and select the location of where the photo is located, follow the on-screen instructions displayed to apply the photo on your page.

Navigating Google Classroom

Once you have created a class, you will be directed to that class' page as shown below.

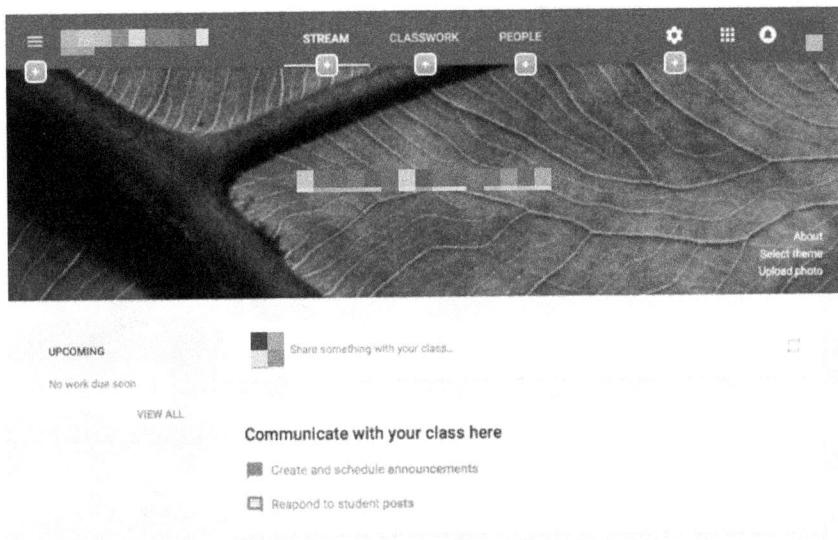

Things to note about this screen:

Menu Tab ▤ : In the Menu tab, other classes can be accessed, as well as your to-do list, calendar, and more.

Stream Tab: The Stream tab is your class' homepage. Here, you will find all of your posts, as well as upcoming work.

Classwork Tab: The Classwork tab is the area for you to create, assign, and grade all the work for your class.

People Tab: In this tab, you can view and communicate both with the students and teachers in your class.

Class Settings ⚙: In the Class settings menu, details about your class, which includes its section, room, and class code, can be found. Also, you can control the ability of students to post and comment on the Stream.

Upcoming: Here, you will find work that would soon be due and other reminders for things such as tests and trips.

Stream (share something with your class...): In the **Stream**, you can post announcements for your class to see. Likewise, you can allow your students to post and comment on the Stream.

Adding Students to Your Class

Via Email

Once your class has been created, you will need to grant your students access to it. One way to do this is to invite your students via email. To do this, first navigate to the **People** tab.

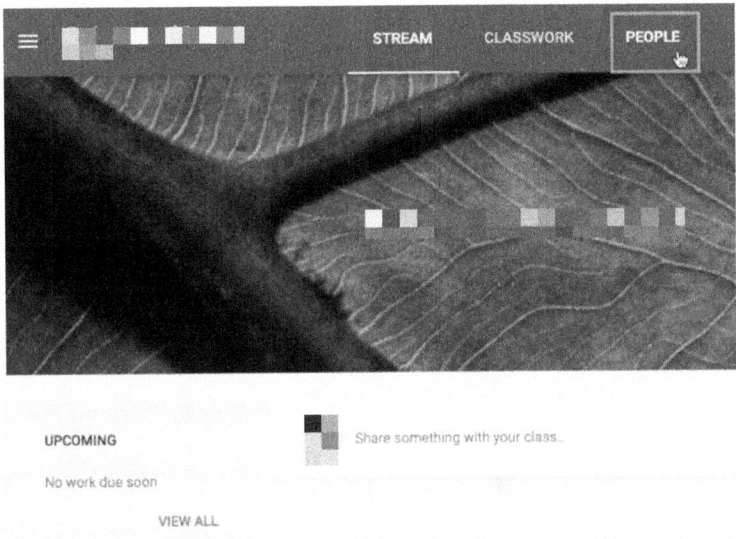

Click the **Invite students** icon.

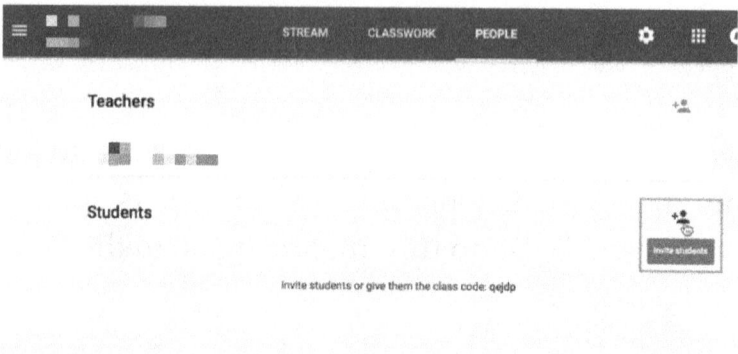

A menu will pop up for you to type your student's email addresses. Once you have added the emails, click **Invite**.

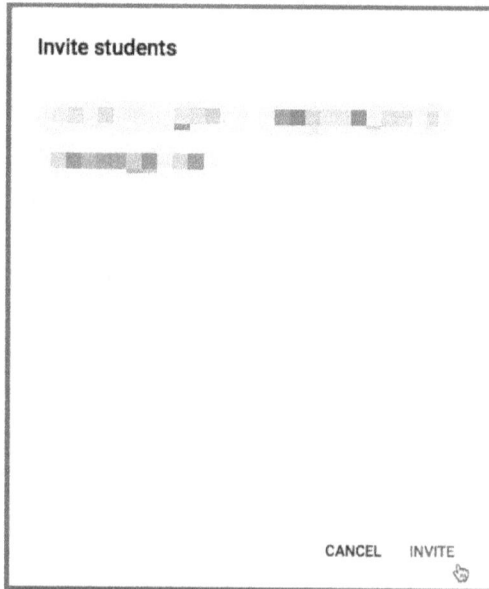

This will send out an email to your students, with a link for them to join your online Classroom.

Via a Class Code

The second option for you to add students to your class is via a **class code** that anyone can use in joining your class as long as you provide them with it.

To add students via a class code, click the **Class settings** icon in the top-right corner.

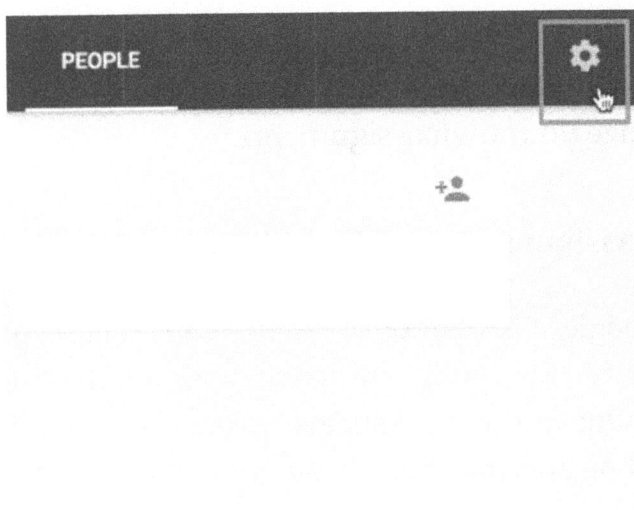

Here, the class code can be found under the **General** tab, from where you can then share your class code with your students.

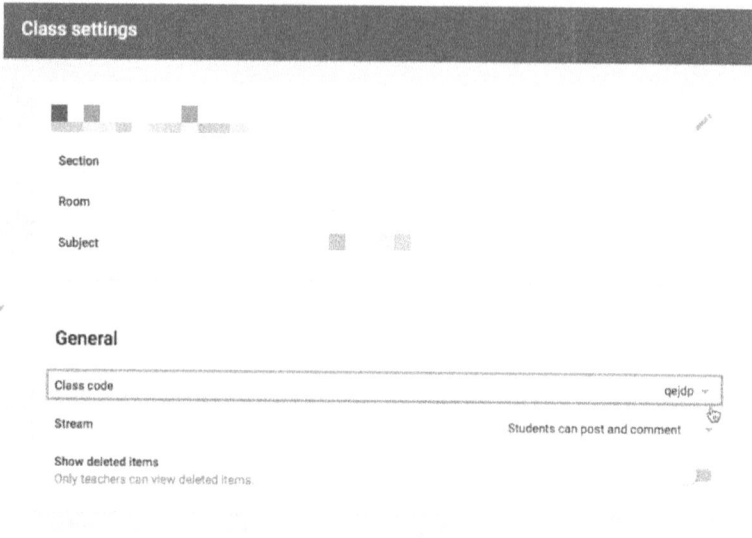

In our **next lesson**, we'll start exploring how to create, track, and organize your own assignments.

Creating Assignments and Materials

With Google Classroom, you can create and assign work activity for your students without having to print anything. Questions, worksheets, essays, and readings can be shared online and made easily accessible to your class.

Creating an Assignment

To create assignments, questions, or materials, navigate to the **Classwork** tab.

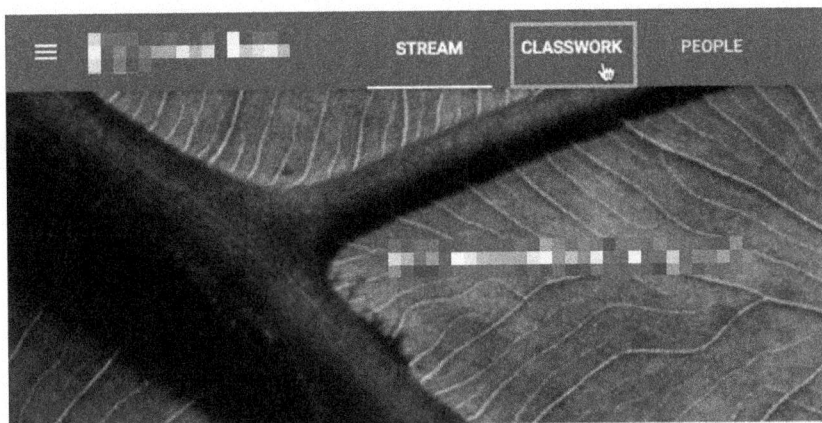

Here, you can create assignments as well as view all current and past assignments. Click the **Create** button to create an assignment, then select **Assignment**. If you would like to pose a question to your students, select **Question**, or **Material** if you just want to post a visual, reading, or other supplementary material.

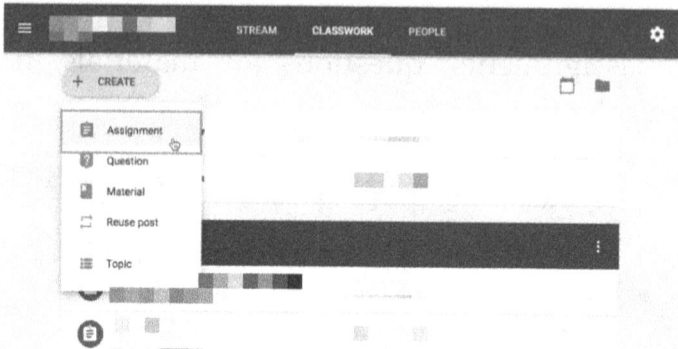

Selecting **Assignment,** for instance, will display the Assignment form. Google Classroom provides considerable flexibility and options when assignments are to be created.

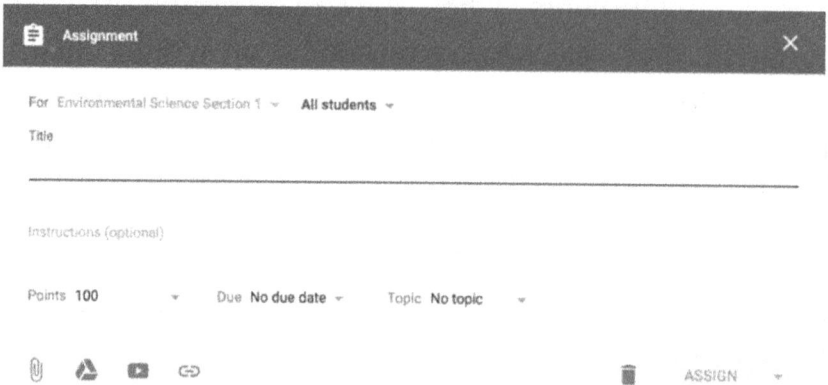

Things to note about this screen

All students: Here, you can send assignments to all students, individual students, or a selected number of students.

Title: Here is where you will type the title of the assignment being created.

Instructions (optional): You can include instructions with your assignment in this field.

Points: This is where you specify the points an assignment is worth. Likewise, if you don't want to grade an assignment, you can specify **Ungraded** by clicking the drop-down arrow.

Due: You can specify the **due date** students must have submitted their assignment by clicking the drop-down arrow and selecting a date from the calendar that pops up.

Topic: Here, assignments and materials can be sorted into topics. You can also select an existing topic or create a new topic to place an assignment under.

📎 △ ▣ 🔗 : This is where files can be attached from your computer, such as files from Google Drive, YouTube videos, and URLs, all of which can be attached to your assignments.

Assign: Click **Assign** once you are satisfied with the assignment created. You can also schedule an assignment using the drop-down menu if you would like it posted at a later date.

Once the form is completed and you have clicked **Assign**, your students will be notified via email informing them about the assignment.

Automatically, Google Classroom takes all of your assignments and adds them to your Google Calendar. Click the **calendar** icon from the **Classwork** tab to show your assignments and to have a better view of the timeline for your assignment's due dates.

Using Google Docs with Assignments

Often times, when assignments are being created, there may be times you would like to attach a document from Google Docs.

This is useful when you want to provide lengthy instructions, study guides, and other materials.

When attaching such types of files, you want to ensure that you choose the correct setting for how your students would interact with it. After attaching a file from Google Docs to an assignment, a drop-down menu with three options would be displayed.

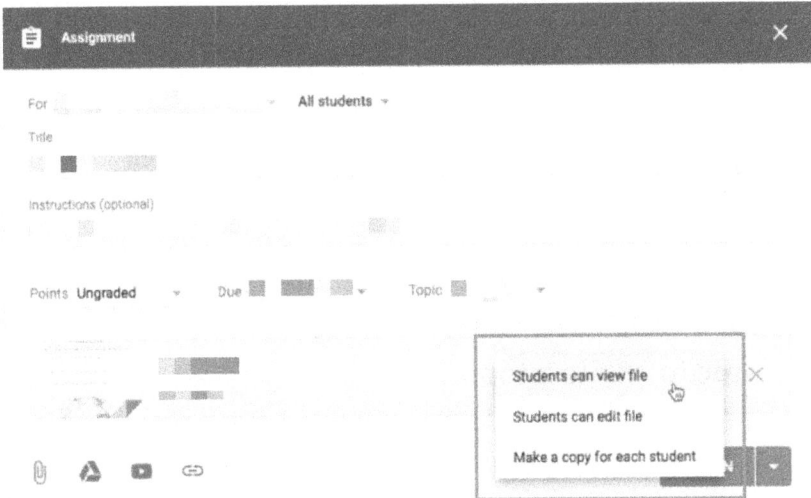

Let's have a look at when you might want to use each of the options:

- **Students can view file**: This option is to be selected if you want your students to view the file without making any changes to it.

- **Students can edit file**: This option is useful if the documents you are providing requires collaboration from the students or if you want them to fill out the document collectively.

- **Make a copy for each student**: Select this option if the document is to be completed by each student. The same document for every student will be created if you choose this option.

Using Topics

Topics can be used to sort and group your assignments and material. From the **Classwork** tab, click **Create**, then select **Topic** to create a topic.

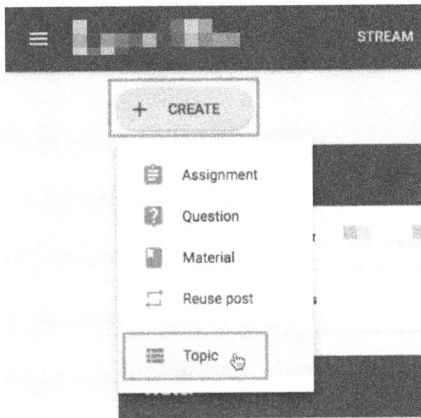

Topics are useful when organizing your content into several units to be taught throughout the year. It can also be used to distinguish your content by type, splitting it into homework, readings, classwork, and other topic areas.

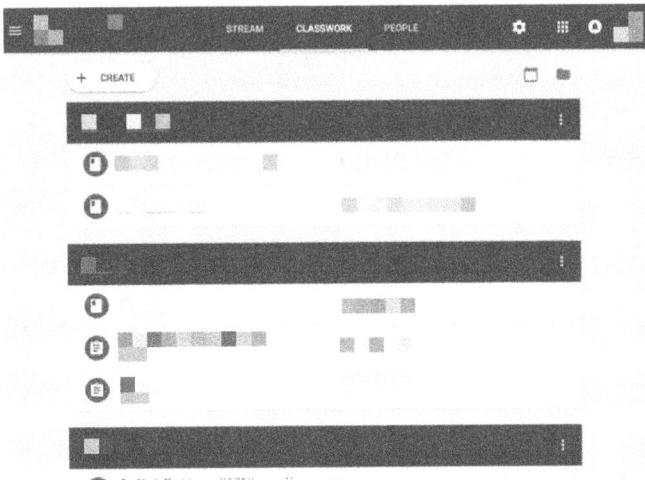

Sorry, I had to blur out most of the information. They are quite sensitive, kindly indulge me.

Using Google Forms with Google Classroom

Typically, Google Forms are used to create surveys, feedback forms, sign-ups, and more. **Quizzes** can also be created with Google Forms, which can be easily be incorporated into Google Classroom. There are several question types that you can use in making your own quizzes, which also features an array of customizable settings.

Creating a Quiz

First, you will need to create a basic form to create a quiz. Please refer in detail here, <u>lesson on creating simple forms</u>, or use the web address <u>https://bit.ly/30x0h2U</u>, but some of the basics will be covered below.

Navigate to the <u>Google Forms homepage</u>, or use the web address <u>https://bit.ly/2WHRnOX</u>, then click the **Blank** icon.

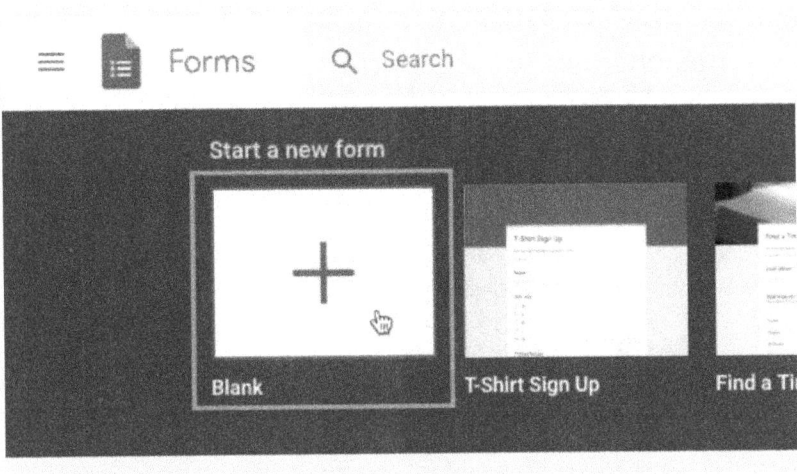

Please note that in some instances, clicking the above links may not show you the screen above but instead may take you to a form that is automatically created for you, all you need to do is make necessary edits as given below;

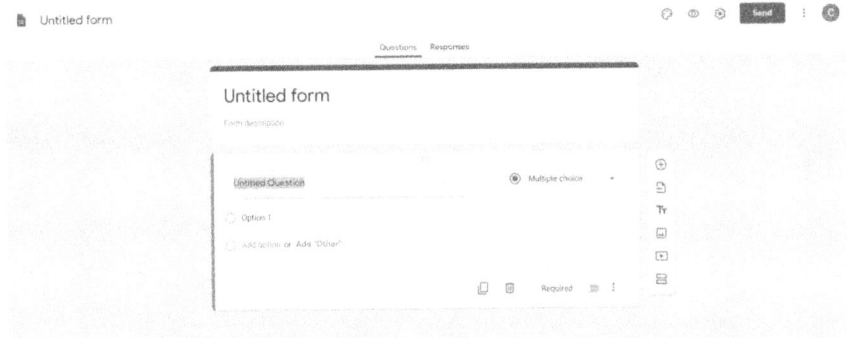

Before you start to write your quiz's questions and answers, make some changes to the form's settings. In the top-right corner, click the **Settings** icon.

Click the **Quizzes** tab and toggle the switch, **Make this a quiz**. Turning this switch on will enable several **quiz options** that will allow you to select how your students are to interact with your quiz.

Settings

GENERAL   PRESENTATION   **QUIZZES**

Make this a quiz
Assign point values to questions and allow auto-grading.

Quiz options

Release grade:

◉ Immediately after each submission

◯ Later, after manual review
Turns on email collection

Respondent can see:

☑ Missed questions ❓

☑ Correct answers ❓

☑ Point values ❓

CANCEL   SAVE

When you are done creating your quiz, you can then attach the quiz to your assignment, and send it out to your students

Things to note about this screen

Release Grade

Immediately after each submission: Selecting this option will ensure that students immediately receive their grade after the completion of the quiz. This option may be used if your quiz is entirely a multiple choice quiz.

Later, after manual review: This allows you to review each quiz manually before releasing the grades to your students. This can be useful for quizzes featuring questions that involve a lot of typed responses.

Respondents can see

Missed questions: Selecting this option will allow students to view questions missed upon completing the quiz.

Correct answers: This will allow your students to view the correct answers to the questions upon receiving their grades.

Point values: This will display the total points for the quiz to the students as well as the number of points they earned for each question.

Once your chosen settings have been selected, click **Save**. After this, you can then name your quiz and begin to write your questions. For more information on creating form questions, click <u>lesson on creating simple forms</u>, or use the web address <u>https://bit.ly/30x0h2U</u>.

Creating a Timed Quiz in Google Classroom

Timed quizzes can be delivered by using a combination of Google Forms, the scheduling function in Google Classroom, and Form Limiter – Google Forms add on. To set a timed quiz, first, create your quiz in Google Forms as earlier described, then install and activate the <u>Form Limiter add-on</u> for Google Forms. Date and time limit should be enabled in the Form Limiter add-on

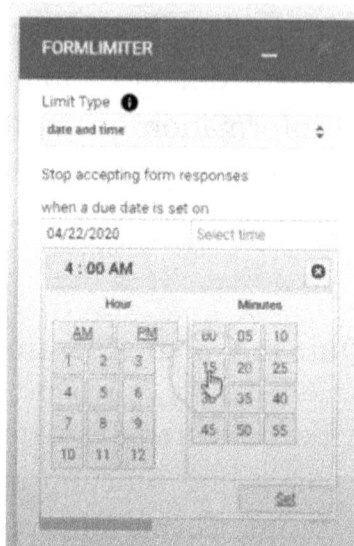

Thereafter, attach your quiz to an assignment, and make it to go live at a particular date and time by using the scheduling tool in the assignment area of Google Classroom. Refer to the section *How can assignments be scheduled in Google Classroom?* for details on how to use the scheduling tool.

Selecting the Right Answers for Your Questions

You will need to specify the right answers for each question on your quiz. To do this, click **Answer Key**.

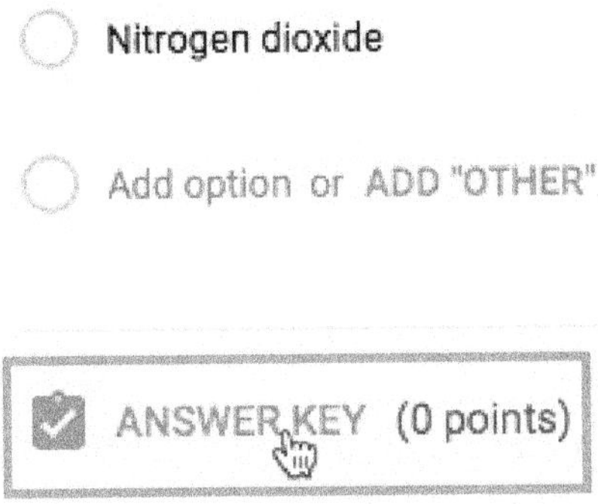

○ Nitrogen dioxide

○ Add option or ADD "OTHER"

☑ ANSWER KEY (0 points)

This screen will display differently depending on the type of question you have selected. Let's take a look at how you can select the right answers for various popular types of questions:

- Using **multiple-choice** or **checkbox** question: Simply select the right answer(s) from choices made available.

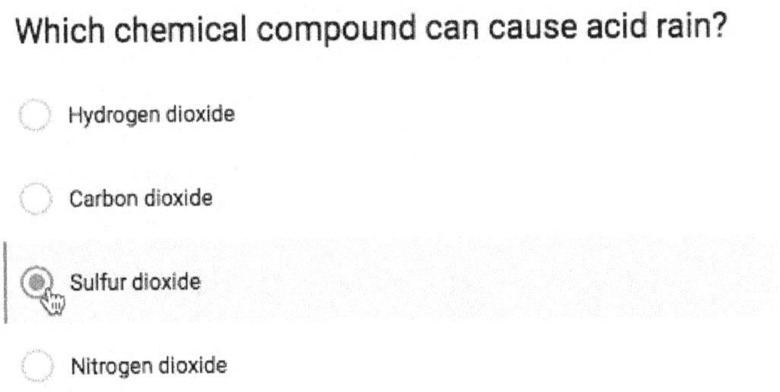

Which chemical compound can cause acid rain?

○ Hydrogen dioxide

○ Carbon dioxide

◉ Sulfur dioxide

○ Nitrogen dioxide

- Using **short-answer** question: Type the right answer in the field, **Add a correct answer**. Also, multiple answers can be added if the wording varies for a particular question. If **Mark all other answers incorrect** is checked, all answers that do not match the correct answer will be marked incorrect. But if left unchecked, you will manually review and grade any answers that are not an exact match.

What term is defined as "the settling of substances ▪ ▪ ▪ ▪ ▪ ▪ ▪ ?

0 : points

Sedimentation ✕

Sediment ✕

Add a correct answer

☐ Mark all other answers incorrect

- Using **paragraph** questions: Paragraph questions do not provide the ability to add the right answers. Because they are longer with more analysis required, you will have to read and grade them individually.

After the right answers have been chosen, select the number of points you want each question to be using the points field.

What term is defined as

10 ↕ points

Also, there is the option of using **answer feedback**. This provides students with feedback on certain questions which depends on if they chose the correct answer. Click **Add answer feedback** and type what you would like as a message to be displayed for correct or incorrect answers, and save.

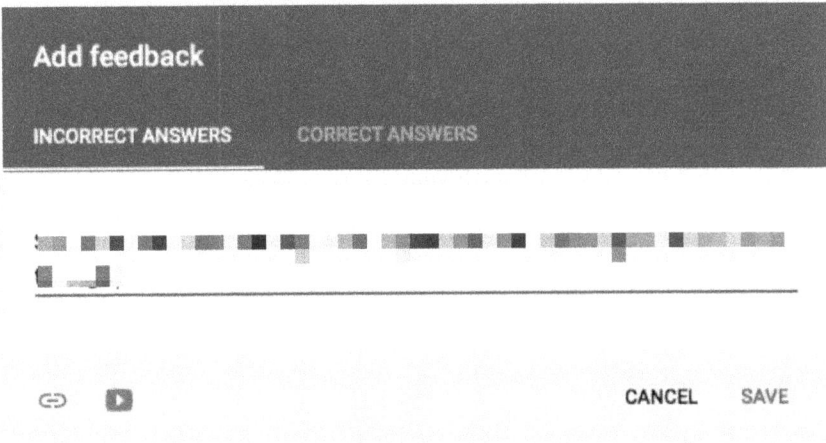

Add feedback

INCORRECT ANSWERS   CORRECT ANSWERS

CANCEL   SAVE

Once you are okay with the question, preview it and test it out to ensure everything works fine as intended. To do test it out, click the **Preview** icon in the top-right corner.

Adding Sections to Your Quiz

You may separate your quiz into several sections depending on how many questions you made available. This will help break up your questions across multiple pages instead of them appearing on one page.

To do this, simply click the **Add section** icon in the toolbar.

Questions can be added to these sections by following the same instructions provided above. Also, questions can be moved to other sections by dragging and dropping them, as shown below.

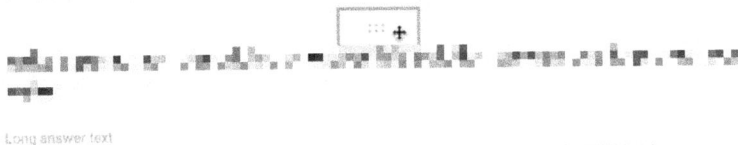

Long answer text

Adding Quizzes to Classes

After creating your quiz, you will need to import it to your class. Forms can be attached just like when attaching documents, videos, and links to your assignments.

To add a quiz to your class, click the **Google Drive** icon, as shown below, at the point of creating an assignment.

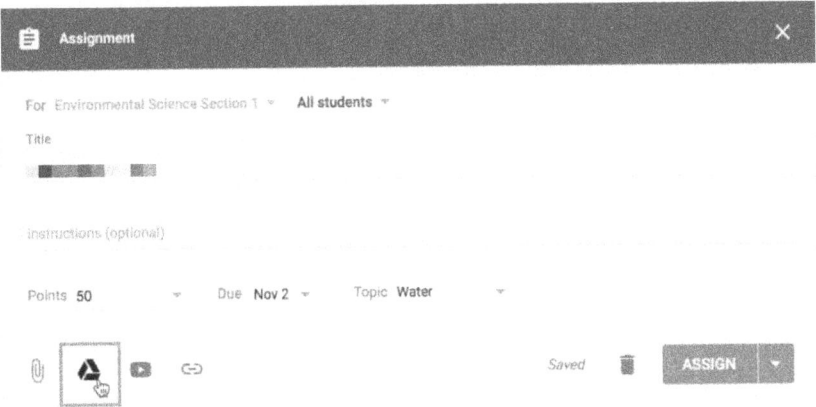

Then locate and select the quiz you have created, click **Add**.

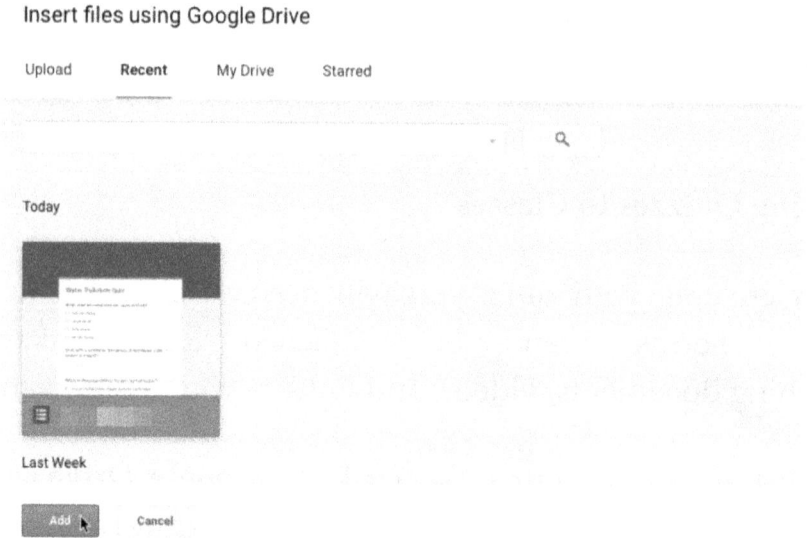

Insert files using Google Drive

Upload   Recent   My Drive   Starred

Today

Last Week

Add   Cancel

After the assignment with the quiz attached has been sent, your students will then be able to complete it. You can access responses using Google Sheets or via https://bit.ly/2ON14XO, just like the way you would with any Google Form.

Grading and Leaving Feedback

After your assignment has been responded to and submitted by your students, you will now be able to review and grade them. Google Classroom provides each assignment with its own page, thereby making it easy for you to grade and leave feedback for your students.

Viewing Individual Assignment

To view an assignment, first, you will have to navigate to the **Classwork** tab. Click the assignment you wish to grade, then click **View Assignment**.

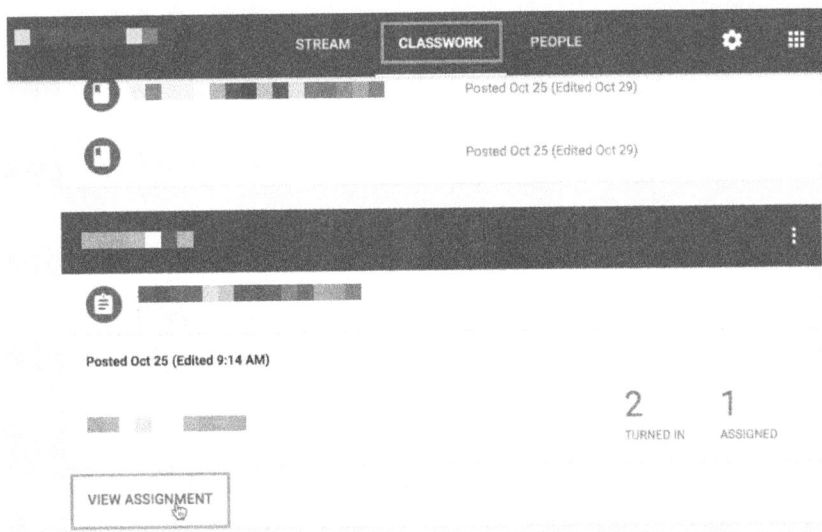

Doing the above will bring up the **Student Work** page for that particular assignment. This is where you will be able to view your student's submissions individually and grade them.

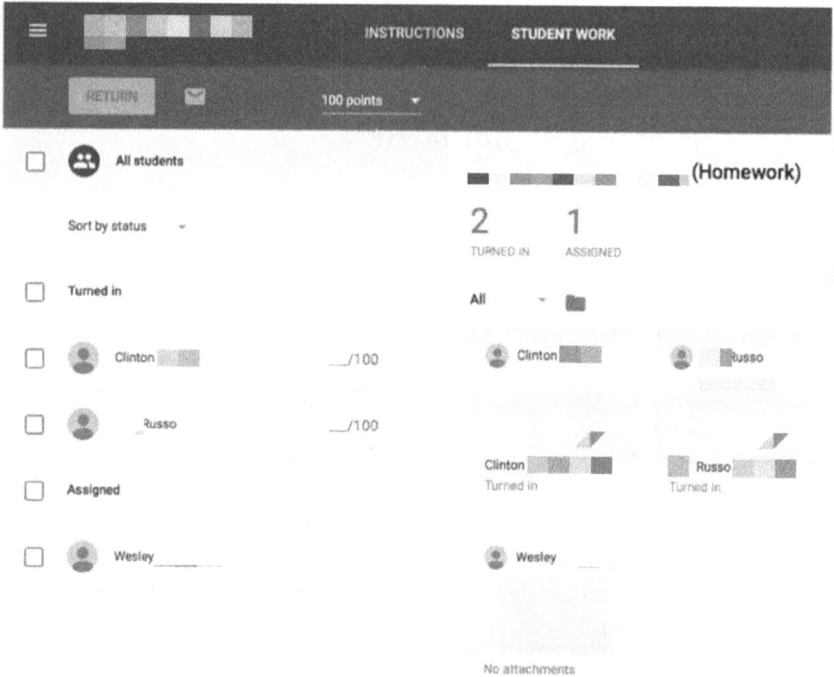

Things to note about this screen:

Instructions tab: In the Instructions tab, you can view the instructions for the assignment you're viewing

Points: This is the total point value for the assignment. You can adjust it by clicking it and typing how many points you'd like the assignment to be worth

Turned in: Here, you'll find a list of the students who have submitted the assignment

Assigned: Here, you will find students that have yet to submit their assignment

Grade point: Click the points automatically assigned to each student that have turned in their assignment, and you would be able to type the grade you'd like to give each assignment. Once you've graded an assignment, it will automatically be selected. You can then choose to return it to the student

All: Here, you will find all of the assignments submitted by your class so far. Simply click one of them to open and view it.

Grading Assignments From Student's Work Page

To grade student's assignments on the **Student Work** page, simply click the score next to the student's name, then enter the grade you would like to give.

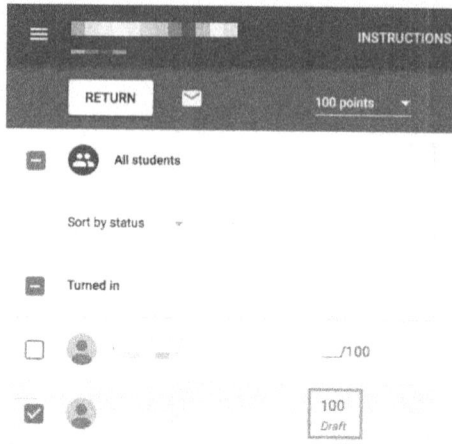

Upon grading the assignments, select them and click **Return** to revert them back to the corresponding students.

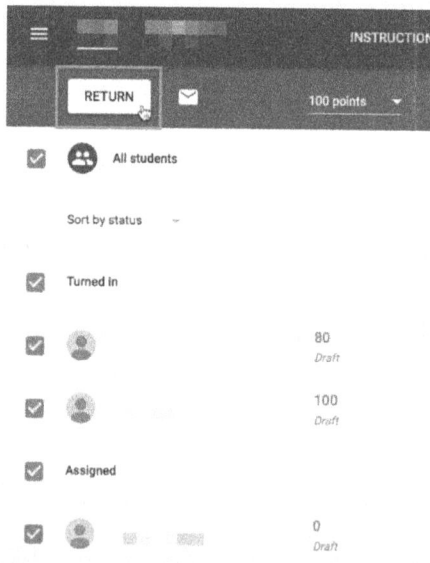

Grading Assignments with Grading Tool

An assignment can also be graded using the grading tool located in the individual submission. First, you will have to click an assignment to open it.

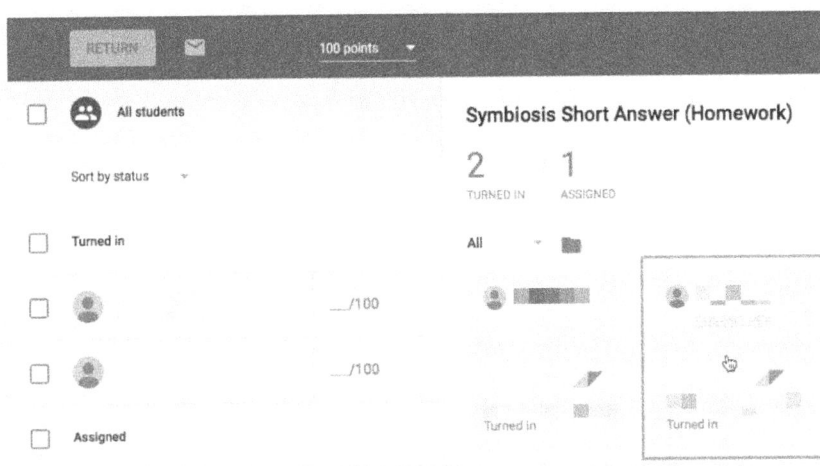

On the right-hand side, there is a column that shows the grading tool. In the **Grade** field, type the grade you wish to give. You can also provide feedback for the student in the **Private comments** field.

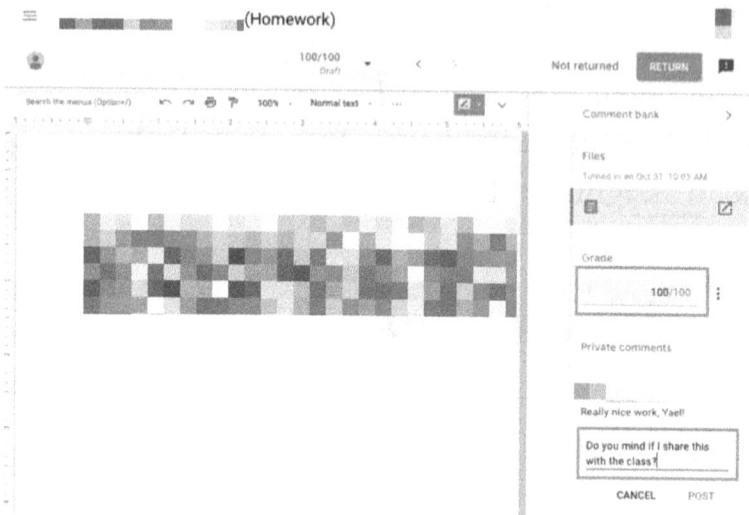

Once you are done grading an assignment, and you are ready to share it with that particular student, click the **Return** button.

Viewing Class Grades As a Whole

With Google Classroom, grades can be exported to Google Sheets from assignments. This will create a spreadsheet that shows the grades for the individual assignment for each student, along with the average grade of the assignment and the class' overall average grade.

To export the grades for the assignments you have graded, go to any **Student Work** page, click the **gear** icon located at the top right-hand side, and then click **Copy all grades to Google Sheets**.

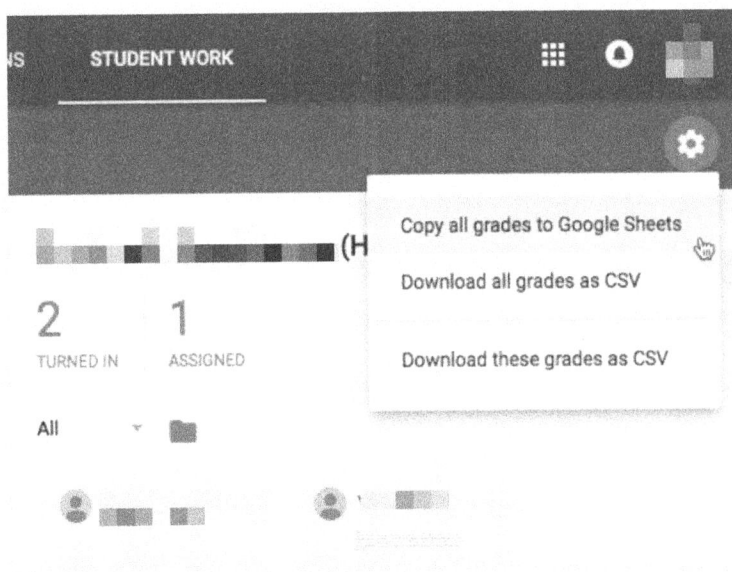

Once one of these spreadsheets have been created, note that it will not automatically update. Each time you grade more assignments, you will have to export the grades again.

If you want a more customizable and detailed gradebook experience, there are several third-party apps that can work in tandem with Google Classroom. A list of such can be found on this page or via this web address https://bit.ly/30vIf0Q.

Communicating with Students and Parents

Google Classroom provides you with the ability to communicate with students via email. Also, it is possible for guardians and parents to receive email summaries to keep them abreast of current and future events.

Emailing Your Students

You can email your students all at once or individually. Irrespective of who you wish to email, first, you will have to navigate to the **People** tab.

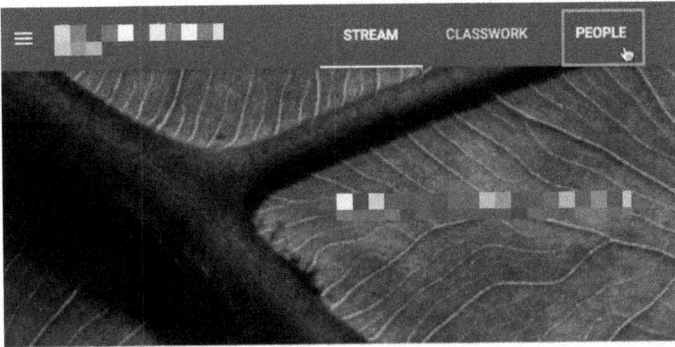

Here, there are several options at your disposal, depending on the people you want to email. Let's have a look at a few possible examples:

- **Emailing a single student**: Locate the student's name, click the **More** button, and select **Email student**

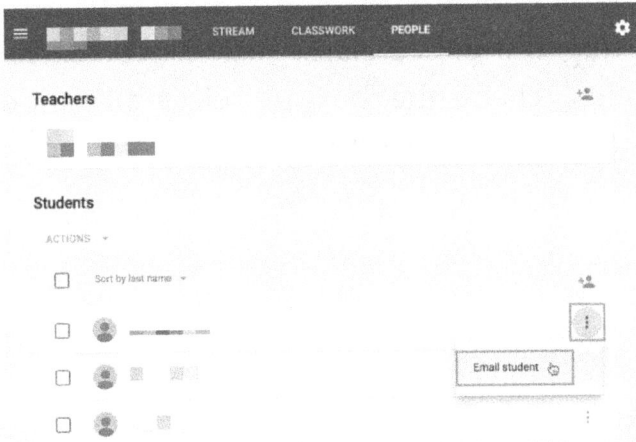

- **Emailing multiple students**: Check the box beside the names of the students you wish to email, click **Actions**, and select **Email**.

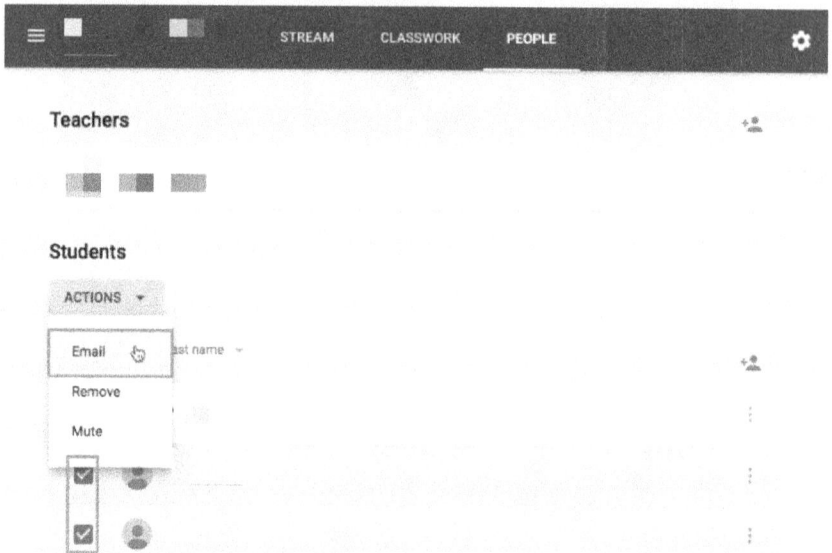

- **Emailing the entire class**: Just above the list of students, check the box to select all students, click **Actions** and select **Email**.

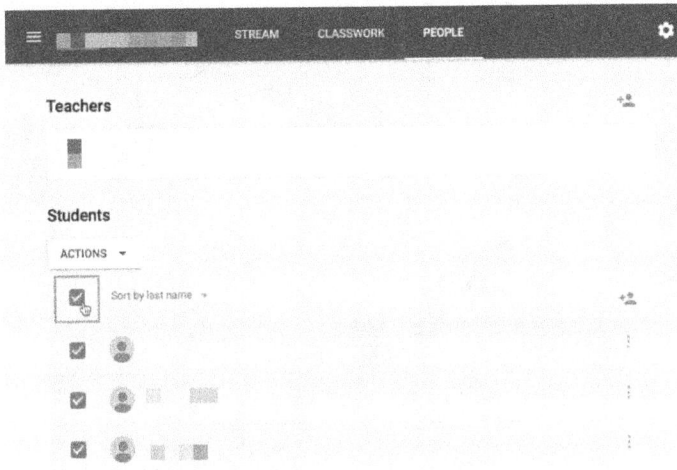

Posting Announcements

Announcements can be posted to the class Stream instead of sending emails. Announcements are posts your students will see on the Stream immediately they are signed into Google Classroom. This can be useful to serve as reminders, notifications of future events, or anything else you would like to share with your students.

Navigate to the **Stream** tab to post an announcement, then click the **Share something with your class...** field.

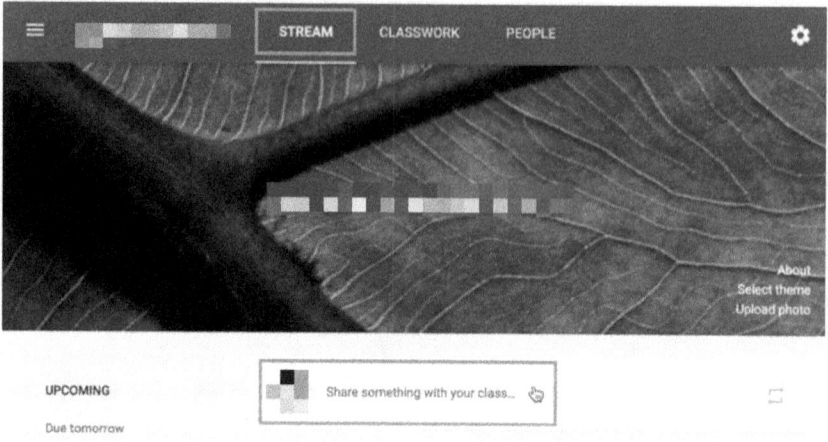

From here, type whatever you wish, and include an attachment, or link if you desire. Click **Post** when you are done.

Class Summaries for Parents

As earlier mentioned, guardians and parents of your students can also receive email summaries, which include missing work, recent class activity, and upcoming work. However, it does not include features that allow for direct messaging with families or for families to comment on their children's work. Class summaries are automatically generated and cannot be modified with additional content or personal messages. Parents can choose if they want to receive these emails on a daily or weekly basis, and as well, they have the option to unsubscribe.

Note: For class summaries to be available for use, your students must be using **G Suite for Education** school accounts. School administrators must also give teachers access to manage the email summaries.

If your students are making use of **G Suite for Education** school accounts and you would like to provide email summaries, click this link or use this web address https://bit.ly/2ZR4teM to learn how to do so.

Student's Guide

Joining a Class

To join a class, you first need to ensure you have signed into your Google personal account, which applies if Google Classroom is used outside the school setting, i.e., for homeschooling purposes. If you do not have one, kindly click this link creating a Google account or use this web address https://bit.ly/32L3RJv for a review lesson on how to open a google account.

However, if Google Classroom is used within a school setting, you would need a school account to sign in to Google Classroom – also called G Suite for Education account since this type of account is most appropriate for a school setting. A school account is set up by the IT Admin department of accredited schools. Your school account could look like **you@yourschool.edu**.

Via a Class Code

After signing into your Google account, go to classroom.google.com, and click "**+**" located on the top right-hand side of the page, then click Join class.

Thereafter, enter the Class code provided to you by your teacher, and click Join.

Join class

Ask your teacher for the class code, then enter it here.

Class code

Cancel   Join

Via Teacher's Invite

You will receive an email invitation if your teacher invites you to join a class. The class can be seen on your Classes page. To join a class via an invite, go to classroom.google.com, and ensure to sign in with your correct google account if you are not yet signed in. If you are signed in already, and you need to switch to the appropriate Google account, click your profile picture in the top right-hand corner and select or add your account. After signing into Google Classroom with your Google account, click **Join** or **Accept** on your class page to access the class.

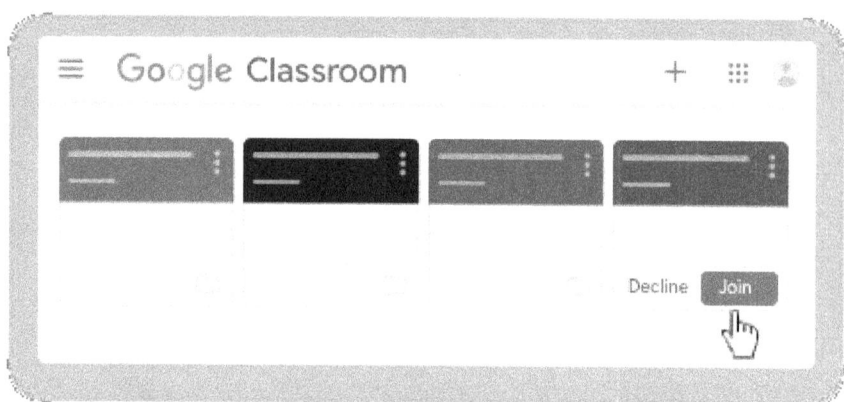

Accessing Announcements and Posting Class Comments Via the Stream

The **Stream** is where students view and access announcements, discussion topics, assignments, and comments. When activated by the teacher, posts and comments can be added by students.

Adding a Student Post to the Stream

Below are directions for students.

1. After joining a class, Google Classroom, for instance, will open the class created by your teacher, let's say, English Classroom, and it will display all instructions, reminders, classwork, and any links to forums and lesson activities for discussion and comments.

2. To add a post to the Stream, click the **Share something with your class** located at the top of the screen.

3. Add the text you want to post to your class. **Remember, your teacher and the whole class can view this.**

4. Use the post to ask questions about class assignments, share important resources on class subject and topics, or collaborate with other students in the class.

5. Other content can also be added to your posts by clicking on the 'Add' tab.

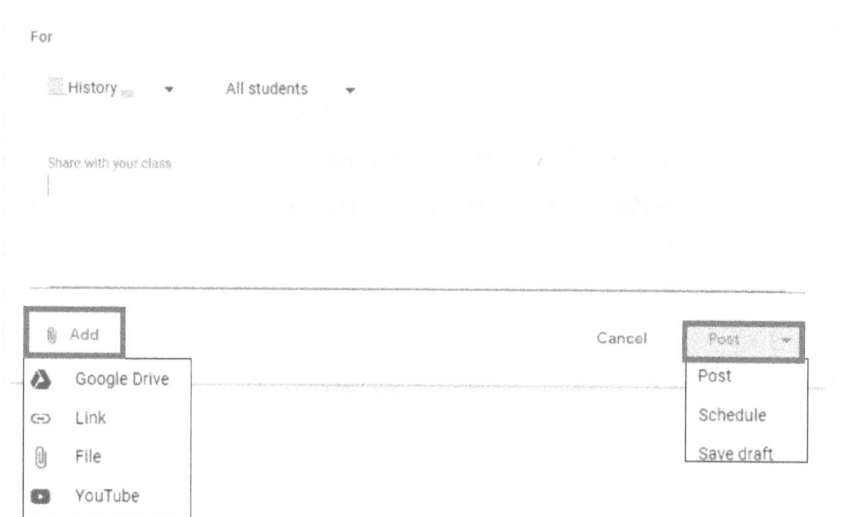

Google Drive: Files saved in your Google Drive.

Web Links to outside websites, resources, etc.

File Attachments: files saved locally on your computer or device

YouTube videos

6. Once your comment, question, attached file, Youtube clip, or link has been added to the class, click the Post tab. (Yes students can also share resources to the class).

Locating and Submitting Assignments

The Classwork tab is where all assignments from your teacher or work to be done are located.

1. To identify the assignments and/ or work that needs to be completed, click the **Classwork** tab.

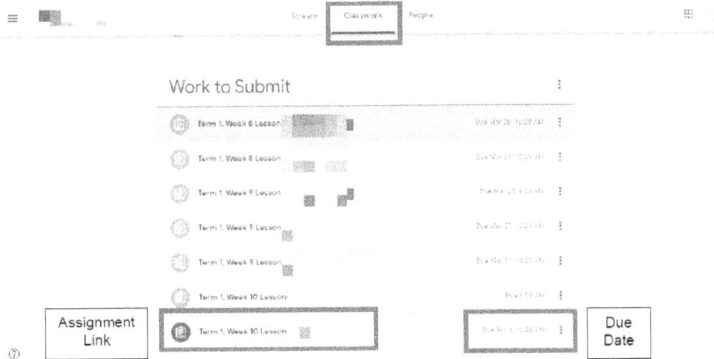

2. To open the latest assignment for instructions, click the latest ⊞ assignment link.

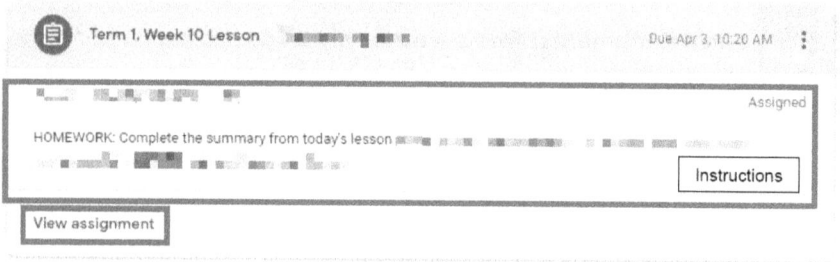

3. Click the **View Assignment**, and the following options will be shown.

① **+ Add or create- Dropdown menu** -Upload a file from your documents or Google Drive or create a doc, slide, sheet or drawing to complete and attach your assignment

② **Private Message**- Teacher can only see.

③ Title, Description of the Assignment and Due Date.

④ Files attached/ uploaded for the assignment.

⑤ **Add a Class Comment**: Whole class can see questions and comments.

⑥ The **Turn In** button when clicked, submits the work to the teacher.

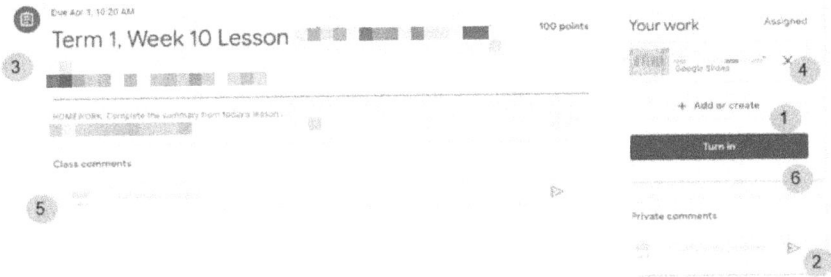

4. When the **Turn In** button is clicked, you will be prompted by a pop-up to confirm your submission.

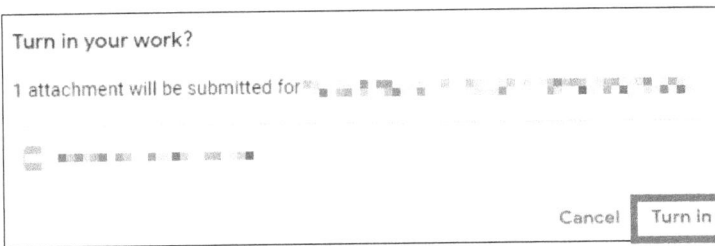

5. Once your work is **Turned in**, a text box will be shown where you can **Unsubmit the work** if you wish to make any changes. However, this can only be done if it is before the due date.

6. Upon making the needed revisions, click **Turn In** again to resubmit the assignment.

Submitting Other Types of Assignments

If the teacher assigns a collaborative Google file, e.g., google doc, the file can be edited by students. In this case, a **Mark As Done** button would be seen instead of **Turn In**. This option will appear only in Google Classroom, and not in the file itself. When assignments have been completed by students, the **Mark**

As Done button is clicked to notify the teacher that they have finished.

If a student is submitting late work, a private comment can be left to inform the teacher of late work submission.

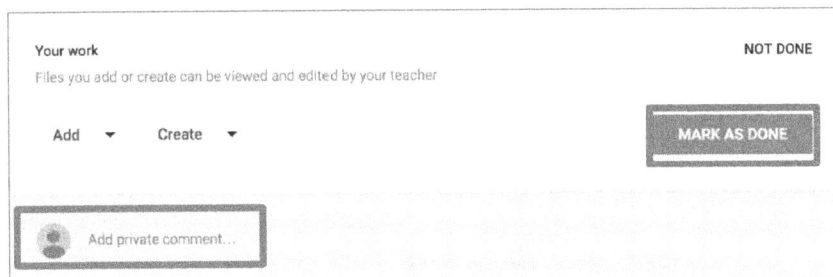

Viewing the List of Assignments for All Classes

Students can view the list of assignments for all their classes by navigating to the **To-Do** page. Go to the Google Classroom Menu homepage (just after signing into Google Classroom), and at the top left-hand corner of your screen, click the 3 lines ☰, and select **To do.**

There are two tabs at the top of this page:

To-Do: This is where students can see a list of all their pending assignments. Click the assignment name to view the assignment's details page.

Done: This is where students can see a list of completed assignments **turned in** or **marked as done**.

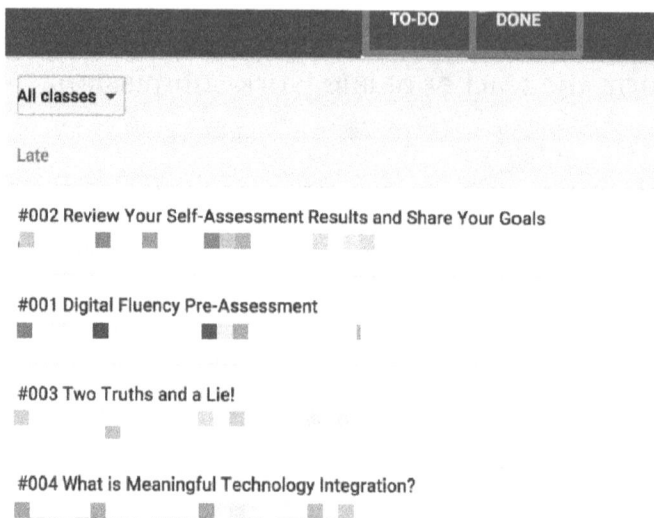

Using Google Classroom Calendars to View Assignment's Due Dates

Google Classroom comes with an integration of Google Calendar, making it easy for students to view assignment due dates and more in a single location. Assignment or discussion questions with a due date will automatically show up in that class's Google Calendar.

Go to the Google Classroom Menu homepage (just after signing into Google Classroom), and at the top left-hand corner of your screen, click the 3 lines ▤, and Select **Calendar**. Here, you will be able to view a weekly calendar of all your classes, or filter for each class. Also, you can go directly to that assignment's details page by clicking on the assignment.

Accessing Your Classroom Files in Google Drive

Google Classroom automatically creates folders for you in Google Drive. To access your classroom files, go to <u>Google Drive</u> (https://bit.ly/39qK8ju), or click ⠿ > ⌂ located at the top-right hand corner of your screen.

On the next screen, there is a master folder titled, '**Classroom**'

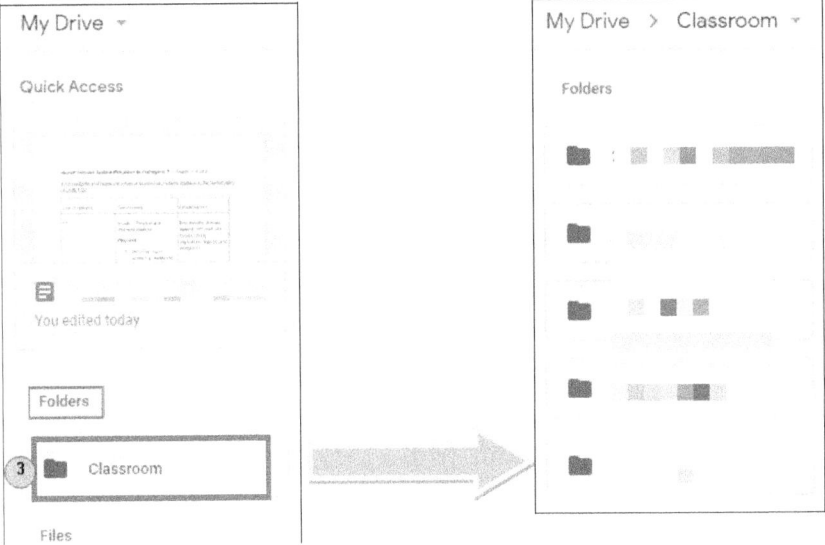

There is a subfolder for each of the classes you have joined, right inside the master Classroom folder, and inside each class folder, there are several files and folders which depends on what your teacher added to your assignments.

A Short message from the Author:

Hey, I hope you are enjoying the book? I would love to hear your thoughts!

Many readers do not know how hard reviews are to come by and how much they help an author.

I would be incredibly grateful if you could take just 60 seconds to write a short review on Amazon, even if it is a few sentences!

\>> Type this web address https://amzn.to/39LJKwe in your browser to leave a quick review

Thanks for the time taken to share your thoughts!

Chapter 3

Setting-up and Using Google Meet for Live Video Classes

For the past couple of years, Google Hangouts, a standalone product, was the main option for video conferencing, for instance, in education, but now that it is gradually being shut down, Google has made Hangouts a part of the G Suite line of products and consisting of two primary products, Google Meet and Google Chat.

This chapter details the step by step guide to enable you get started with setting up online classes using Google Meet, meeting and interacting with your students, and getting on with your teaching in no time.

What Is Google Meet?

Google Meet is a video-conferencing tool that not only doubles as a website and an app for iOS and Android devices but also allows you to join in through a link or phone call with the capacity to present your screen instead of your video feed. With the help of Google Meet, teachers can remotely conduct online classes with students, record the class for future reference, and share the contents of their screen, among many others. If you already use Google Classroom or have got the hang of how Google Classroom works, then teaching online with Google Meet should be a walk in the park and should be pretty seamless for you.

Before now, Google Meet was a premium service that was available only to organizations that uses Google's G Suite line of products. However, that has now changed as it has been made free and open to everyone. However, the free version requires that a free Google account be created, providing you with a maximum of 24 hours meeting time up until Sept. 30, 2020, after which it reverts to its regular 1 hour meeting time for free users.

Why Google Meet is a Must-Have for Teachers

Google Meet is a perfect tool to use for distance learning with so many features that goes with it. Not only can online classes be conducted just like you would in a real classroom, but you can also draw over a whiteboard to better explain things, share the contents of your screen, mute participants to provide a seamless teaching experience, and more. Also, you can pin a student's video feed to better interact with them, chat with students to discuss specific topics, take student's attendance, and divide students into several sections within a classroom.

Estimating the Google Meet Capacity of Your Students

Google Meet, until recently, was available only to G Suite users, but currently, it can be used for free if you have a Google account. However, usage of this tool has a set participant limit per session with 100 participants and up to 250 participants for free and non-free users, respectively. Below is a comprehensive overview of the number of students that can be hosted on Google Meet using your Google account.

Google (or G Suite) account type	Maximum number of students per meeting
Regular (non-G Suite) Google account	100
G Suite Education, G Suite Basic	100
G Suite Essentials, G Suite Business	150
G Suite Enterprise Essentials, G Suite Enterprise, G Suite Enterprise for Education	250

Setting-up Google Meet Via Meet Homepage

After deciding the Google/G Suite account you wish to use, the next step is to proceed in setting up your Google Meet session, i.e., the classroom equivalent for your students. To initiate this process, head over to Google Meet homepage using the web address https://meet.google.com/, then click on "Start a meeting," if you are the teacher or "Join a meeting," if you are the student. You will then be prompted to sign in with your Google/G Suite account if you haven't yet done so, and depending on whether you are using a Google/G Suite account, you may be required to enter the name of your meeting in the dialog box that is displayed. Click **Continue** once you are done, after which you will be asked to grant permission to your camera and microphone.

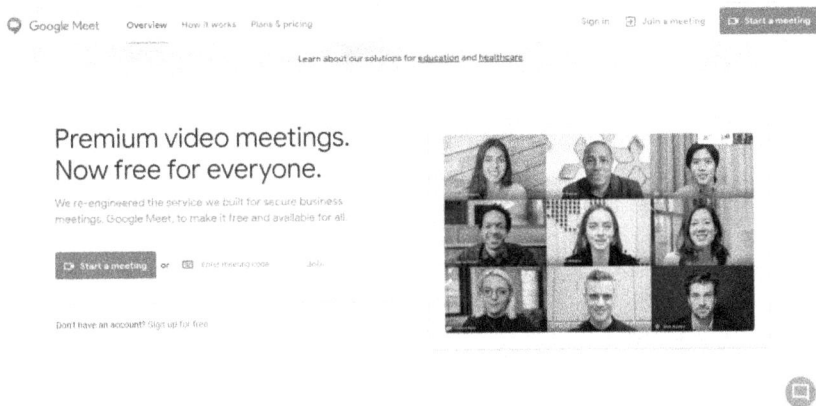

However, if you have previously signed in to your Google/G Suite account, head over to Google Meet homepage, click on "New meeting" to display the details of how you wish to start a meeting such as a meeting link to share with your students to join the virtual class, starting an instant meeting, or scheduling the meeting in Google Calendar. Click "start an instant meeting." Google Meet will now request your permission to access your camera and microphone.

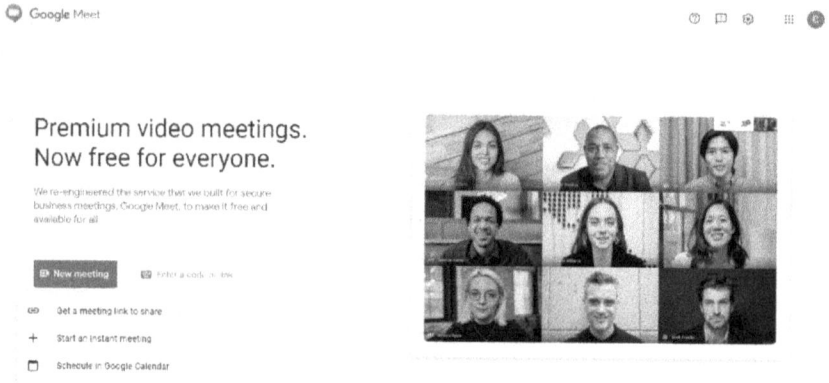

Inviting Students to Join Your Class Via Meet Homepage

After granting the necessary permissions, click **Join now**

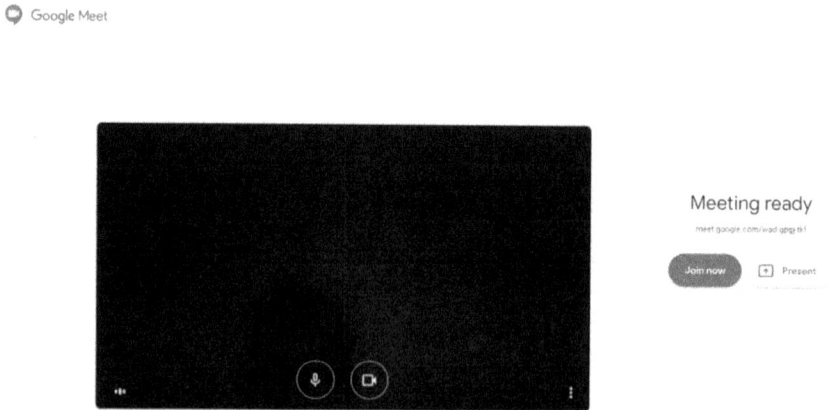

Once you click on **Join**, you will be taken to another page where you will be prompted to add others to your meeting. Click **Copy joining info** from the dialog box that is displayed.

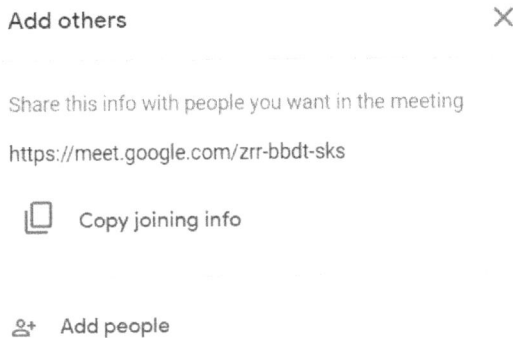

Add others   ✕

Share this info with people you want in the meeting

https://meet.google.com/zrr-bbdt-sks

     Copy joining info

     Add people

Note: Depending on whether you are using a Google/G Suite account and your country of residence, a US telephone number would also be displayed in the dialog box through which your students can call in rather than use the meeting link.

After copying the joining info, head over to your email account, paste, and send the info you copied to your participant's email addresses. They will then be able to join your class via the link provided, or call in using the telephone number.

Note: You can also add students from your contact list on your Google account by clicking **Add people** at the bottom of the dialog box, or preferably, you can share it as a post or announcement in your Google Classroom either as a Material or Assignment.

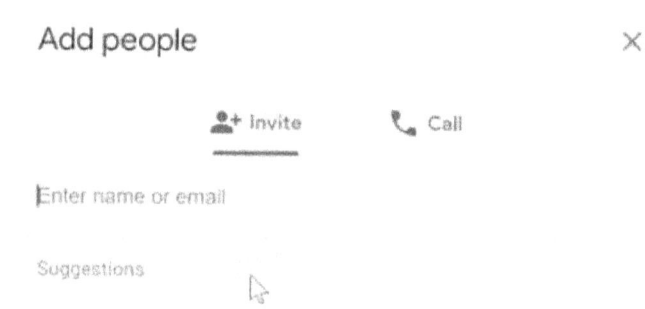

And that's it. You have now successfully sent out invite links to your students, which when clicked, they will be directed to the page where they would join your class for a live video session using the join button as shown below

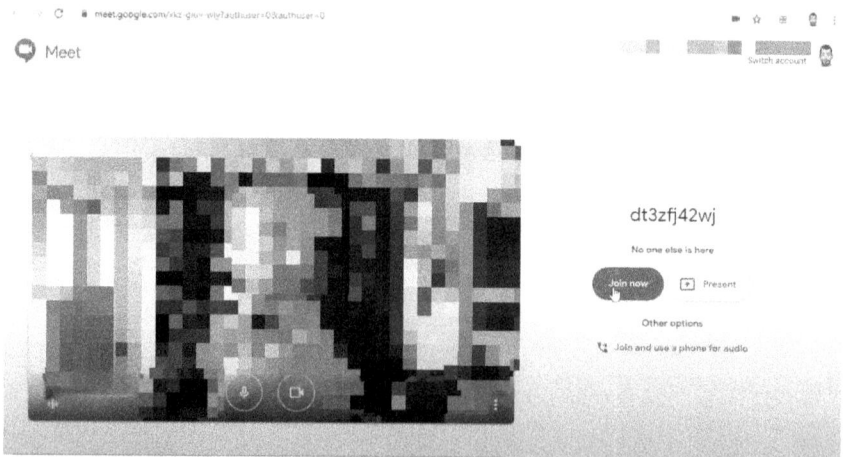

Using Google Meet in Google Classroom

All the above steps discussed are valid for anyone who wishes to use Google Meet for video conferencing irrespective of the type of Google account they have, i.e., personal Google account or G Suite account or if they use Google Classroom or not. However, to make use of Google Meet inside Google Classroom for those who wish to use Google Meet with Classroom, you are required to have a G Suite account, e.g., G Suite for Education. Google Meet is seamlessly integrated into Google Classroom, meaning a unique Google Meet link can be generated for each of your classes and for the students in that classroom. The link for the Meet can either be displayed on the Stream page of your class and/ or in the Classwork page for easy access by you and your students. This link can be used over and over again anytime you want to have a video conference with your students. However, your students can't join the Meet without you, which means they can't access the Meet before you or join after you have left the Meet video conference.

Setting-up Google Meet Via Google Classroom

Without much ado, let's take a look at the specifics in setting-up Google Meet in Google Classroom and putting this new integration to use.

To begin, sign in to google classroom with your G Suite account, e.g., G Suite for Education, and open the class you wish to activate Google Meet for. Click on the gear icon in the top right-hand corner to access the class settings

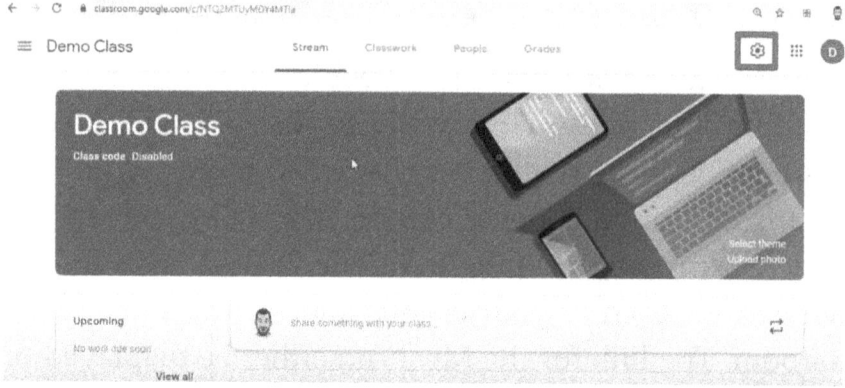

Scroll down to the **General** section of the settings and under **Meet**, click **Generate Meet Link**

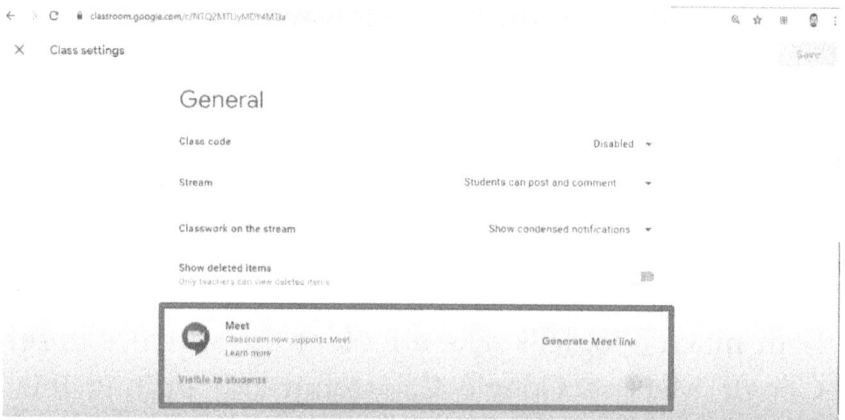

A unique Google Meet link will then be created for your class and your students. However, if you do not want the link to be visible to your students yet, toggle off the **Visible to students** settings. If this setting is turned on, it will be accessible both on the Stream and Classwork page of your class. Also, you can click the drop-down arrow beside the Meet link to either copy the link and share it with someone who is not part of your classroom (e.g., another teacher or guest speaker if needed) or reset the link to generate another unique link (if needed). When you are done, click **Save** on the top right-hand corner to save the changes.

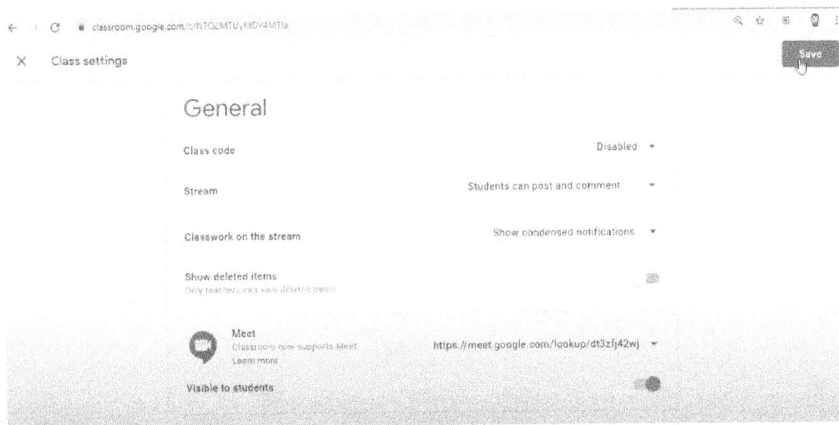

This would make you or your student access the Meet link either on the Stream page and/ or on the Classwork page which when clicked, they will be directed to the page where they would join your class for a live video session using the join button as shown below

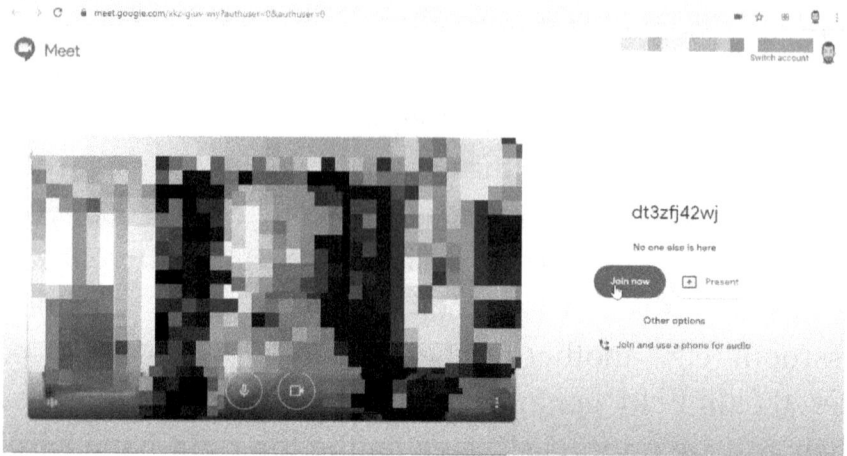

Also, you can share the link as a post or announcement in your Google Classroom, either as a Material or Assignment.

There are other ways to invite students to join your meeting. You can schedule the meeting via the calendar in the Classwork page of Google Classroom. For a description on how to schedule a calendar meeting in Google Classroom, follow this link https://bit.ly/2P46ePj

Sharing Your Screen with Your Students

After setting up Google Meet either via Google Meet homepage or Google Classroom, it is important to share your screen before your students join your class since they can't join your class before you. To share the contents of your screen, click the **Present** button inside your Meet class. This will display 3 options you can choose from to share your screen, which is: share an application window from your computer, your entire screen, or for simplicity, a single chrome tab browser. Select the entire screen, the browser tab, or the application window you want to share, then click the share button.

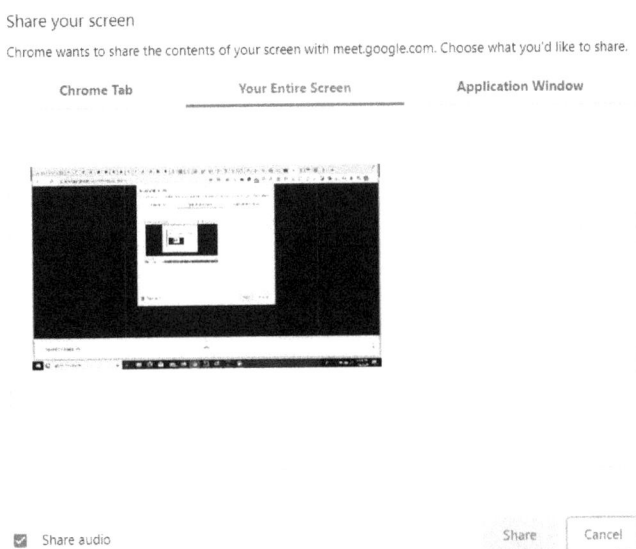

Would you love to take the teaching experience to the next level by using Google Jamboard inside Google Meet in presenting your topic on a whiteboard just like you would in a real classroom? If yes, then click this link or use this web address https://bit.ly/39AeDDC for a step-by-step guide on how a whiteboard can be created inside Google Jamboard and shared on Google Meet.

Taking Attendance of Your Classroom

Once you have set up your classroom and your students have been added to the class, you will need a means of verifying students that are present during a session and those that aren't. For this purpose, you will have to download and install the Google Meet Attendance extension on Google Chrome using the link https://bit.ly/2Eszn4L. This tool will help you in creating a spreadsheet for your classroom by adding the existing student's names and the time they logged in.

To take attendance in Google Meet after successfully installing the tool, do the following:

Step 1: Ensure you are in an ongoing Google Meet session with your students.

Step 2: Lookout for a checkbox tab just beside the **People** tab.

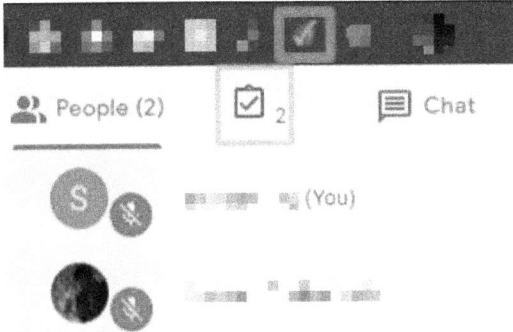

Step 3: Click the tab so that you're taken to the Google Sheet where your student's attendance details are stored.

Step 4: Hover the tab for more options, then click the "+" spreadsheet icon to create a new spreadsheet if, for instance, you have longer meeting sessions. On the other hand, the first horizontal toggle is used for automatically logging in the participants.

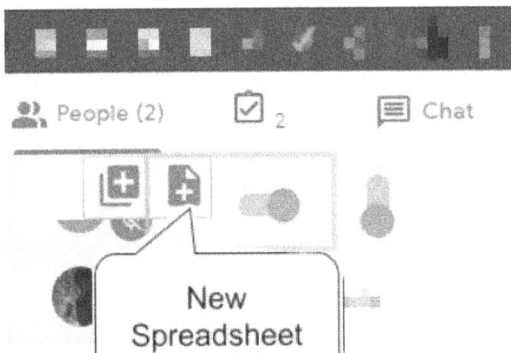

New
Spreadsheet

Step 5: To capture the current attendees, toggle the horizontal icon off and on, then to access the spreadsheet, click the checkbox tab. The name of the attendees (students), the meeting URL, and the time of their joining the meeting would be displayed on the sheet.

Muting Students During a Live Session

Student's microphone can be muted to provide an engaging class session and to prevent any noise interference with an ongoing class. This can be done by selecting the **People** tab, clicking the down-arrow beside a student's name, and then clicking the microphone button to mute their microphone. At the moment, you can't mute your student's microphone all at once on Meet. This would have to be done individually, which could be very stressful. However, this functionality is in development by Google, which may have been deployed at the time of writing this guide, or after.

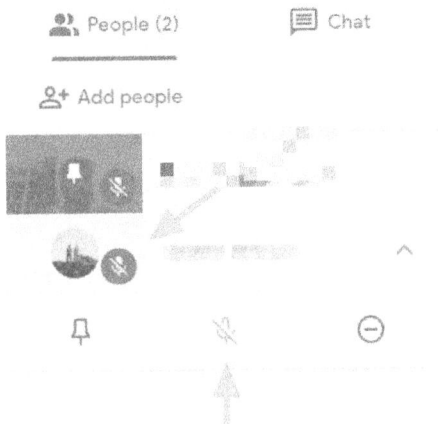

Also, no option is currently available in Google Meet to disable the unmute function. This implies that even after individual participants have been muted, a participant can still unmute themselves. Instead, you will have to request that the muted participant unmute himself/ herself. Google is aware of this problem, and it should be fixed in the coming updates, which may have been deployed at the time of writing this guide, or after.

Looking to create study groups or breakout rooms for your students inside a Google Meet classroom? Click this link or use this web address https://bit.ly/2P3S2Wz for a detailed guide on how this can be achieved.

Recording a Google Meet Video Meeting

Like other video conferencing services on the market such as zoom, Google Meet also has the ability to record your video meetings, which can be saved, and shared for future reference. This makes it possible for **you to** share them with your students that were not able to partake in the meeting, thereby helping them recap what was taught earlier in the day.

Note: The recording feature of Google Meet is available for free only for users with a G Suite Account up until Sept. 30, 2020, after which you would need a paid license to record your classes. However, you would still have access to your recordings even after Sept. 30.

Let's have a look at how you can record your meetings in Google Meet.

What You Need

- Google G Suite account

- Enough free space in your Google Drive

- You must be the meeting organizer or must have a G Suite account in the same organization as the meeting organizer

Starting a Recording

To start a recording in Google Meet is straightforward and does not require much.

Step 1: First, you must have joined a meeting as earlier described, then once you have joined in, click the **3-dot** icon in the bottom right-hand corner, and select **Record Meeting**.

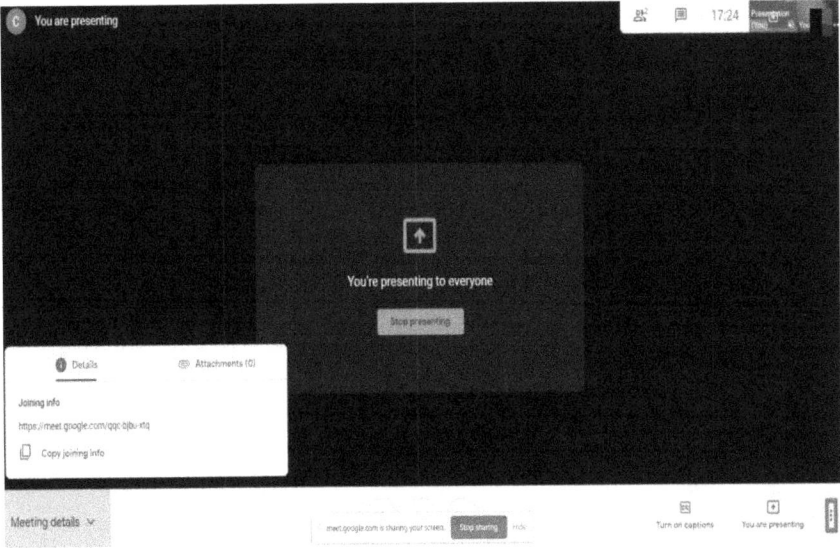

●	Record meeting
■	Change layout
[ ]	Full screen
CC	Turn on captions
⚙	Settings
📞	Use a phone for audio
!	Report a problem

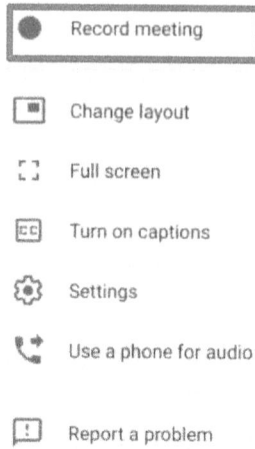

Step 2: A **Ask for consent** dialog box will appear. Confirm your selection by clicking **Accept**. After being confirmed, all meeting members will receive a notification that you have initiated a recording, and that the recording will be started.

Stopping a Recording

To stop a recording in Google Meet is much like starting a recording but in the reverse order.

Step 1: While the meeting is ongoing and being recorded, click the **3-dot** icon in the bottom right-hand corner, and select **Stop Recording**.

Step 2: In the confirmation dialog box that is displayed, confirm your selection by clicking **Stop Recording**.

Where Are the Recordings Saved?

After stopping a recording, meeting recordings would be saved in a folder called **Meet Recordings** in the Google Drive of the meeting organizer. Also, you and the meeting organizer will receive an email with a link to the saved file. Also, a link will be attached to the calendar entry if a calendar event was created for your meeting.

What is Recorded?

Only the active speaker and presentations are recorded in Google Meet. Every other participant is excluded from the recording irrespective of who was pinned to the top of your screen. Also, any windows or notifications that you opened or received during the meeting would not be captured in the recording.

Chapter 4

Must-Have Apps and Extensions for Google Classroom

10 Best Google Classroom Apps and Extensions to Improve Classroom Interaction

Google Classroom-integrated apps may not be new to you, or perhaps, you may wonder what integration even means. Basically, Google Classroom utilizes what is called an API (application program interface) to connect and share resources and information with most apps. There are several of such apps that integrate with Google Classroom to improve the teaching and learning experience for teachers and students. I will highlight a few of the best picks below.

- Google Cast for Education:

  Google Cast for Education is a free Chrome app that permits teachers and students to share their screens remotely from anywhere in the Classroom for resource sharing, discussions, and learning together. Google Cast for Education comes with in-built controls for teachers that allow them to seamlessly add students from Classroom.

- **Kami**

  Kami is a go-to app that integrates beautifully Google Classroom. It is used for annotating on documents and PDF files and is probably the most recommended and sought after PDF editor in the education space used by most schools. It is freemium, which means some of its features can be used for free.

- **BrainPOP**

  BrainPOP, a Google app, allows teachers to directly import their classes from Google Classroom. The integration of BrainPOP with Google Classroom begins with the Google app, accessible by teachers via the Google Apps Launcher menu. Importing a Google class into BrainPOP will automatically create student accounts in BrainPOP, allowing students to log in to BrainPOP via the Google Launcher menu. With BrainPOP, the synced class roster can be updated by teachers by adding or removing students in the Classroom.

- **Khan Academy**

  Khan Academy provides teachers with free personalized tools for learning and data-driven insights. With Khan Academy, teachers can import their Google Classroom rosters and can assign Khan Academy content directly to the students.

- **Actively Learn**

Once class roster has been shared between Classroom, apps like Actively Learn makes it easy for teachers to sync assignments, grades, and student submissions back to Google Classroom. For example, a reading lesson can be assigned by teachers directly from Actively Learn to a class in Google Classroom. Also, after an assignment in Actively Learn has been graded, the grades can then be published to Google Classroom.

- **Explain Everything**

Explain Everything is an interactive whiteboard app in which interactions that occur between teachers and students can be recorded and shared.

- **Math Games**

Math Games integration syncs Google Classroom with MathGames.com, which is a source for math games and online skill practice. With Math Games, math assignments can automatically be created and synced by teachers and can also track student's progress. This is great when teachers want to give homework a break by assigning fun math games and activities to their students.

- Listenwise

  Listenwise is a listening skills platform with a large podcast library of quizzes, lessons, and interactive transcripts that provides advanced literacy and learning to all students by improving their listening and comprehension skills.

- **Science Buddies**

  Science Buddies allows K-12 students and teachers to easily find help and free project ideas in virtually all areas, from physics to food science and microbiology to music, among others. It is packed with free lessons, videos and experiments. Extra assessments and quizzes are available when used with Google Classroom.

- **DuoLingo**

  Most teachers, as well as most governments around the world, see Duolingo as the go-to learning companion for language classrooms. With Duolingo lessons, each student is given personalized feedback and practice, thus helping them prepare in getting the most out of classroom instruction.

Chapter 5

Cool Tips and Tricks to Enhance Productivity with Google Classroom

Most teachers are jumping on board with Google Classroom at the moment, sink or swim!. In such an unprecedented time as this where schools are being shut, Google Classroom has become the go-to platform for communication and distribution of remote learning. In this chapter, I will reveal inside knowledge of Google Classroom that most teachers are not aware of by providing you with specific tips to help you in maximizing the effectiveness of Google Classroom during remote learning.

Using the Mobile App to Annotate and Draw on Documents

Are you aware that there are some features in Google Classroom that can only be used on mobile (iOS and Android)? One such cool mobile features are the ability to annotate and draw on documents, which can be Google Docs, Microsoft Word Docs, PDF documents, JPEG, or GIF files. This feature is great for most types of assignments, allowing teachers to provide meaningful feedback to students. Likewise, this can be used by students to draw a response, label an image, annotate on text, and more. To give it a try, <u>here are the step-by-step</u> <u>directions for iOS and Android devices</u>, or you can use this web address <u>https://bit.ly/3g3qaxS</u> instead.

Randomly Selecting Students Using the Mobile App

Google Classroom comes with the student selector feature that is only available in the mobile app. The student selector helps you to select students from your class roster randomly. This provides you with a great way to call upon students fairly during class. You can also mark a student as absent, or skip a student to be called upon later.

To use this feature on your mobile device, first, tap "Classroom," followed by your class, then "People."

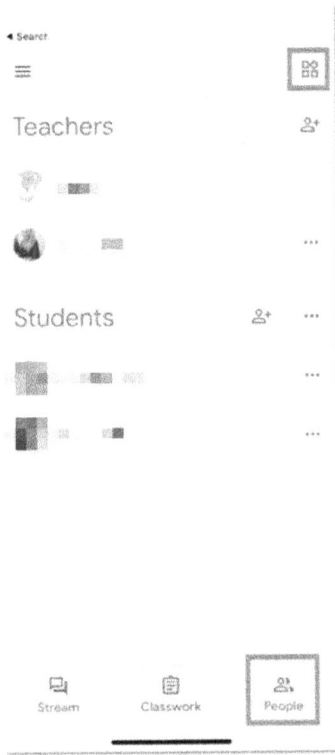

Then click the **Student Selector** icon located at the top right-hand corner and choose one:

- Call upon the student shown ❯ tap Next for another student. You can swipe to the next student on iOS devices.
- Tap Call Later to skip the student shown.
- Tap Absent to mark the shown student as absent.

- To restart the session, tap Start Again (Optional).

To learn more about the Student Selector, click <u>here</u> or visit this web address <u>https://bit.ly/39B2LkZ</u>.

Organizing and Filtering Classwork by Topic

In the Classwork page of Google Classroom, you can create topics to organize your assignments and materials. The Classwork page can be organized in any way that makes sense to you and your students. Most teachers create topics for subjects (suitable for elementary), units, weeks of study, and more. Once topics have been created, a list of such topics would be displayed on the left-hand corner. Clicking on a topic will filter the page to view only the items labeled with that topic.

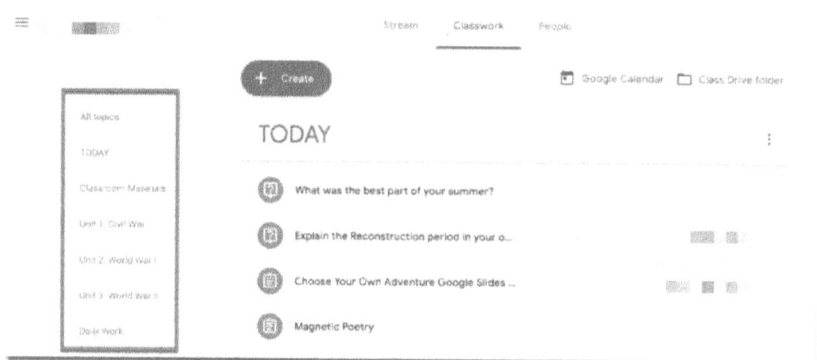

Using Locked Google Forms Quizzes

Google Form Quiz can be locked down to prevent students from opening other tabs to cheat. This is possible for those with managed Chromebooks! With the locked mode enabled, students will be kept "locked" on the Google Form Quiz.

This is how it works:

Important: To use locked mode, you need:

- Chrome OS 75 and up.
- A Chromebook managed by your school for each student.
- A G Suite for Education Account.

When the quiz is in locked mode:

- Students will be unable to open other browser tabs.
- An email is sent to the teacher when a student exits the quiz and reopens it.
- Unmanaged devices would be unable to access the quiz.

When your quiz assignment is created in Google Classroom, beside **Locked mode on Chromebooks**, slide to the right to turn on ⬤.

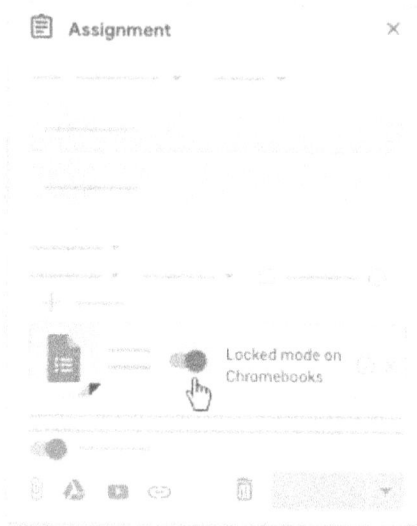

Linking Directly to an Assignment or Topic

In Google Classroom, a direct link to a specific assignment or topic can be gotten. This is very handy when you need to refer students to something specific.

To obtain the link, navigate to the Classwork page and click the 3 dots on the assignment, then click "copy link."

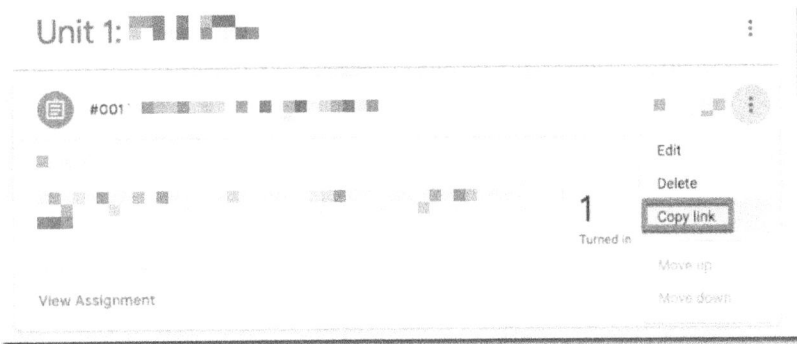

This link can then be sent via email, added in a comment for students, or posted in other platforms or documents. Likewise, you can copy the link of an entire topic. Simply go to the topic on the Classwork page, click the 3 dots, then click "copy link."

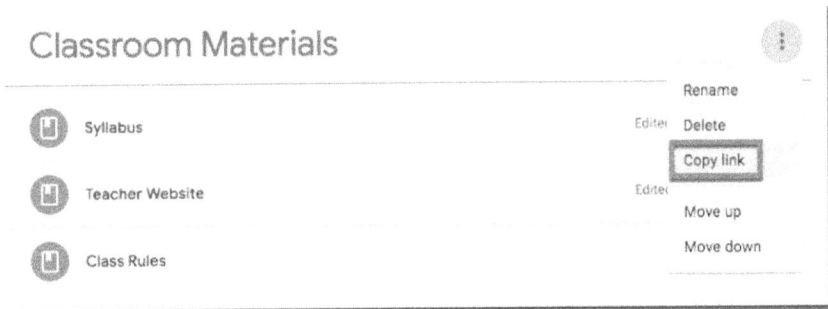

Tagging Students in Class Comments

Tagging specific students in comments is a great way to communicate and provide feedback to students. There are two types of comments in Google Classroom; class comments and private comments. When students are tagged in class comments, you can communicate and answer questions to specific students, which is particularly useful if you allowed students to comment and/or post in Google Classroom.

To tag a student in a comment, type the "+" sign followed by the student's email address, then your comment.

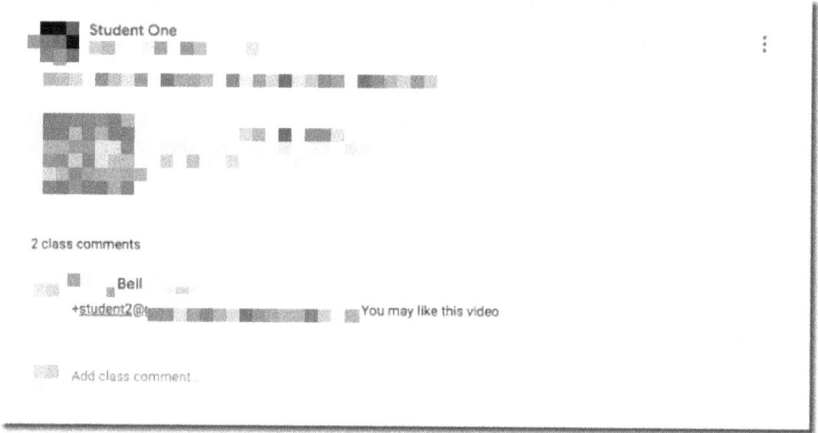

Customizing Stream Notifications for Classwork Page

As earlier mentioned, the Stream is used for communication purposes such as announcements to keep students informed of any activity.

By default, the Stream shows notifications of new activity or items posted on the classwork page. This may feel like duplication for some teachers and can easily muddy up the Stream. On the Stream page, you can choose an expanded or collapsed view for Classwork notifications. You can as well hide them completely from the Stream page.

To do this, click the particular class you want to apply this setting to after logging into classroom.google.com. Click settings on the Stream page, then select an option under General, next to "Classwork on the stream."

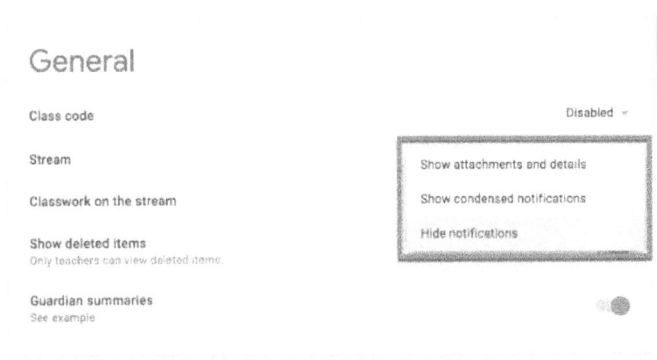

General

Class code	Disabled
Stream	Show attachments and details
Classwork on the stream	Show condensed notifications
Show deleted items	Hide notifications
Only teachers can view deleted items.	
Guardian summaries	
See example	

Bookmarking the Classwork Page

Sometimes, navigating your way around Google Classroom can be tiresome, frustrating, and can take quite too many clicks to get to your desired location. Since you will spend most of your time on the Classwork page, I recommend that you bookmark it.

Bookmarks can be added to your browser's bar for easy access. To turn on the bookmarks bar, for instance, in the chrome browser, click the 3 dots located at the top right-hand corner, then select bookmarks. Ensure the bookmarks bar is enabled, then go to the Classwork page for the class you want to bookmark. Click and drag the lock beside the URL to the bookmark bar to add the classwork page to your bookmark.

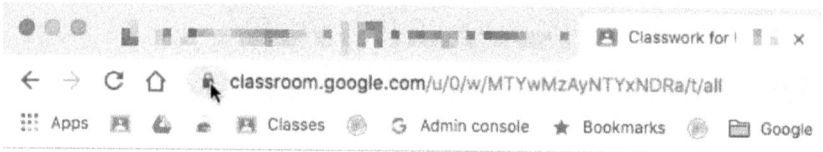

Right-click on the bookmark to shorten or edit the name.

The end... almost!

Hey! We've made it to the final chapter of this book, and I hope you've enjoyed it so far.

If you have not done so yet, I would be incredibly thankful if you could take just a minute to leave a quick review on Amazon

Reviews are not easy to come by, and as an independent author with a little marketing budget, I rely on you, my readers, to leave a short review on Amazon.

Even if it is just a sentence or two!

So if you really enjoyed this book, please...

>> Type this web address https://amzn.to/39LJKwe in your browser to leave a quick review on Amazon.

I truly appreciate your effort to leave your review, as it truly makes a huge difference.

Chapter 6

Google Classroom Frequently Asked Questions

I decided to put together a list of the most frequently asked Google Classroom questions after discovering a trend in the same questions being asked over and over again, questions that several teachers have been searching for answers to. So, I hope this page is useful to educators who are looking for help with using Google Classroom and using it effectively. Let's have a look at some of these FAQs.

How can assignments be distributed to multiple classes at once?

If the same subject is taught across several class periods, you can create an assignment and have it distributed to several classes at once, thus saving you some valuable time.

When a new assignment is created in Google Classroom, click the drop-down arrow beside the name of your class, then use the checkbox to select the classes you wish to distribute the assignment to.

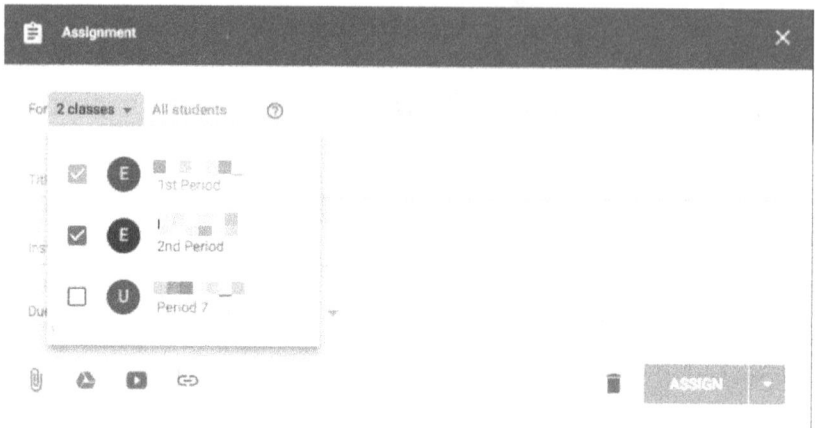

Notes:

- If the assignment is edited, each assignment would have to be edited individually in each class.
- Posting to individual students over several classes at a go is not possible. Instead, when you post to multiple classes, such posts are shared with every student in each class.

How can assignments be scheduled in Google Classroom?

There are times you may need to create assignments before time, which students shouldn't see in their feed until the time is appropriate for them to see it. To schedule an assignment in advance, first, create your assignment, then on the bottom right-hand side, click the drop-down arrow and select **schedule**. Select the date and time the assignment is to be posted on, then click **schedule**.

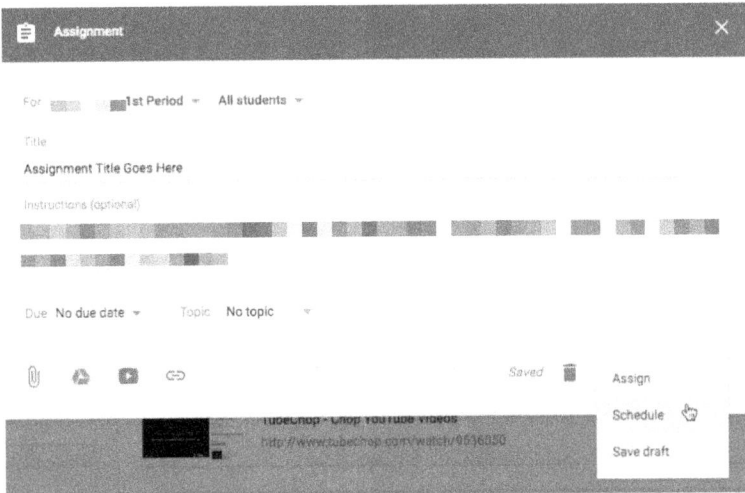

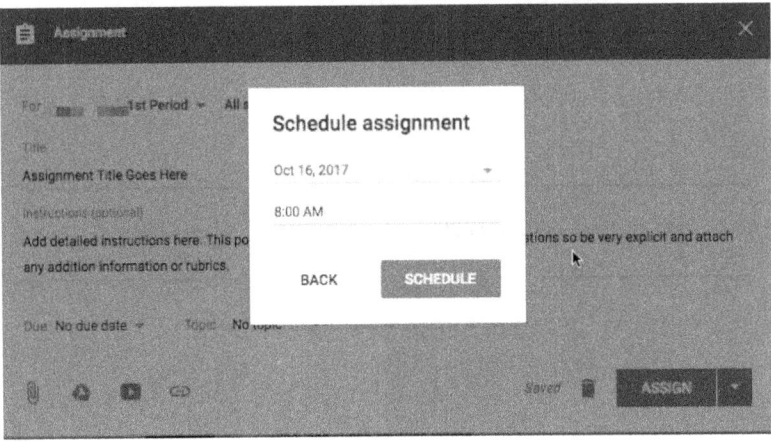

Can assignments be scheduled to multiple classes at a go?

Assignments or posts can only be scheduled one at a time to a class. To schedule for several classes, the process just discussed above would have to be followed for each class. However, you can, at once, save as a draft for several classes by selecting the checkbox for the individual class.

Can assignments expire or be hidden in Google Classroom?

Assignments cannot be hidden, or made to expire. If a student turns in an assignment late, it will show as late, and you will see the assignment was turned in late. Also, since assignments cannot be hidden, your best bet is to schedule the assignments, or instead, leave them as a draft until when you want to post them. Again, this cannot be done after you have already posted the assignments. This can only be done when first creating the assignments.

How can work be assigned to individuals and groups?

Assignments can be assigned to a particular student or a group of students instead of the entire students. By default, all students are made to receive assignments. To change this setting, click the drop-down arrow beside "All Students," deselect "All Students," and select only the student or group of students you want to send the assignment to.

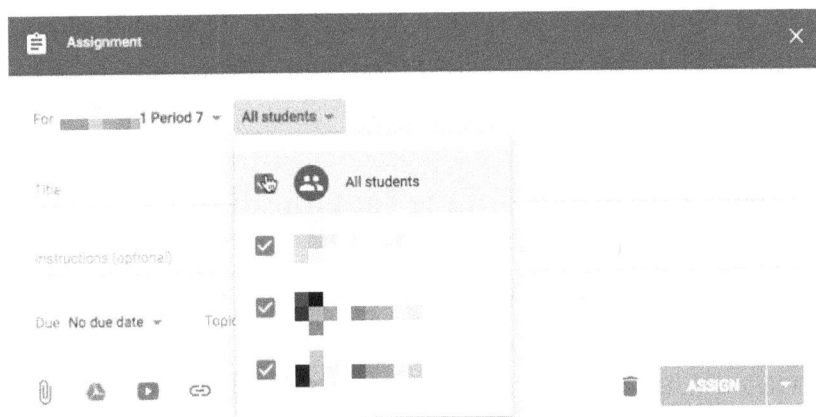

What is the relationship between Google Classroom related and G Suite for Education?

Google Classroom is a product of G Suite for Education, just like Google Docs, Google Sheets, and Google Slides. A user of Google Classroom, as part of G Suite for Education, can make use of other applications, or products such as Google Docs, Sheets, etc. This allows teachers and instructors to collaborate and communicate better with students.

Do teachers and students need to have a Google account to access Google Classroom?

If teachers and students are using Google Classroom within a school setting, then they can access Google Classroom using their school's G suite education account without the need for a Google account. But if accessed outside a school setting for personal and homeschooling purposes, then they would need to have a Google account.

Who is eligible for the G Suite for Education?

K-12 (a terminology that is used in the US and Canada for school grades publicly supported from Kindergarten to grade 12) institutions and higher-education institutions that adhere to the following criteria below.

- Institutions that have a verified non-profit status
- Government-recognized, and formally accredited institutions that provide an internationally or nationally approved certifications at the primary, secondary, and tertiary level.
- Homeschools are likewise eligible if verified by the state homeschool organization.

Do students need an internet connection to work on assignments?

No. Students can work offline on Google docs, Google Sheets, and Google Slides by selecting the "Make Recent Files Offline" option.

Conclusion

I'd like to thank and congratulate you for transiting the lines of this book from start to finish.

I hope this book helped in providing you with a clearer understanding of what Google Classroom is and how important this tool can be for both teachers and students in providing a seamless, virtual and online distance learning. In Chapter 2 of this book, I described with graphical illustrations how teachers can put this great tool to work and how to navigate the Google Classroom environment in creating and grading assignments, quizzes, and also communicating with students and teachers. I also explained how students could likewise make use of this tool to find and submit assignments and post comments to the Stream. Chapter 3 focused on how you can set up live video classes for your students using Google Meet as well as navigating your way around the usage of this tool. In Chapter 4, I discussed the 10 best apps and extensions that integrate with Google Classroom for enhancing the learning experience as well as provide a more teacher-student engagement. This is to assist you in making better choices when trying to decide what apps and extensions to utilize for your online classes. Chapter 5 elaborated on the tips and tricks to help you in maximizing the effectiveness of Google Classroom. And Chapter 6 addressed the common questions and problems asked and encountered by teachers. This is to help ensure that you have a 360-degree insight to effectively putting Google Classroom to maximum and productive use.

At this point onward, you are now better equipped to take on Google Classroom to continue providing class lessons to your students irrespective of location, a tool that has, in recent times, dominated the online distance learning space, and that has won the hearts of many teachers, students, and school administrators. Therefore, it is my utmost desire that you found this book to be quite an interesting read, and helpful as you embark on the journey of virtual learning.

Thank you for reading.